Introduction to Information-Based High-Tech Services

For a listing of recent titles in the *Artech House Technology Management and Professional Development Library,* turn to the back of this book.

Introduction to Information-Based High-Tech Services

Eric Viardot

Artech House
Boston • London

Library of Congress Cataloging-in-Publication Data
Viardot, Eric.
Introduction to information-based high-tech services / Eric Viardot.
p. cm. — (Artech House technology management library)
Includes bibliographical reference and index.
ISBN 0-89006-647-7 (alk. paper)
1. Computer technical support. 2. High technology industries—Customer services. I. Title. II. Series.
QA76.9.T43V53 1999
621.39'068'8—dc21 99-30788
CIP

British Library Cataloguing in Publication Data
Viardot, Eric
Introduction to information-based high-tech services. —
(Artech House technology management library)
1. High technology services industries — Management
2. Electronic information resources — Marketing
I. Title
338.4'7'621'39
ISBN 0-89006-647-7

Cover design by Lynda Fishbourne

685 Canton Street
Norwood MA 02062

International Standard Book Number: 0-89006-647-7
Library of Congress Catalog Card Number: 99-30788

10 9 8 7 6 5 4 3 2 1

Contents

Acknowledgments

THIS BOOK IS DEDICATED to the executives and managers at the following high-tech services firms who agreed to share with me their professional experience.

ANDERSEN CONSULTING
AMERICA ONLINE
ATOS
CAP GEMINI
COMPAQ
EDS
EXPERIAN
HEWLETT-PACKARD
IBM GLOBAL SERVICES
STERIA
SYSECA

Introduction

A REVOLUTION IS AT WORK in the high-technology industry: the irresistible growth of high-tech services. Consider the case of IBM. In 1983 hardware revenues represented 83% of its total turnover, while its service revenues were only a meager 2%, three times less than the software revenues. In 1997 services contributed to 25% of the total revenues, while the hardware represented only 46%. During the period, services revenues have grown from $0.8 to $19.3 billion, meaning a 25% annual growth rate (see Figure 1). IBM Global Services, a new division created in 1997 and already the leader in computer services, has locations in 163 countries and employs 120,000 people.

IBM is leading the pack of firms that are offering a new range of sophisticated services to their corporate customers, different from the traditional hardware maintenance and repairs. Some firms are computer manufacturers like IBM, Compaq, or Hewlett-Packard, others are consulting firms like Andersen Consulting, and still others are information

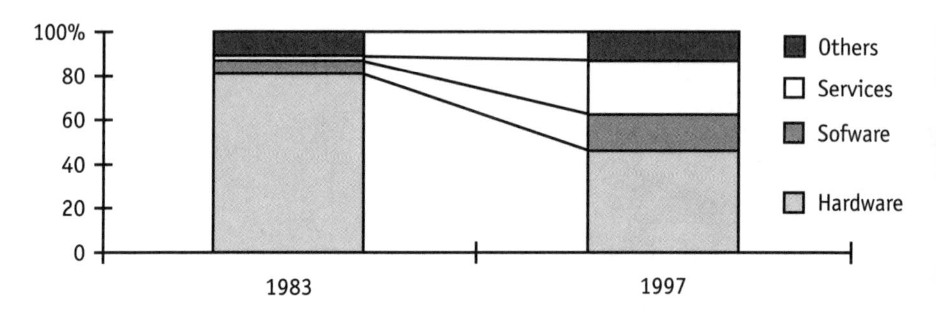

Figure 1 IBM: Evolution of service revenues.

services companies like the American EDS, the French Cap Gemini, or the German Debis.

Answering the needs of their customers and building on the infinite capacity of information technologies, those companies are reinventing the notion of service and offer what we call "high-tech services," a subtle mix of information, knowledge, and technology.

This trend may seem to be limited to business-to-business activities, but with the explosion of new telecommunications media such as direct broadcast satellites, Global Mobile Personal Communication Services (GPMPCS) like Iridium, cellular and wireless data, private networks, and of course, the Internet, it looks as though these high-tech services are reaching consumers as well.

Among the latest communications media, the Internet is the one that has been experiencing the most phenomenal growth—more than any other computer application through late 1993 and 1994. During this short period of time, usership has grown exponentially. It is difficult to say just how large the Internet is.

The number of commercial World Wide Web (WWW) sites has expanded from 370 in June 1994 to 50,000 in January 1996. Finally, the intensity of traffic and transactions over the Internet WWW (estimated at under $500 million, at best) is spreading continually. For example, International Data Corporation (IDC), a specialized consulting firm, expects the total value of goods and services purchased via the WWW this year to surpass $10 billion, with business-to-business transactions accounting for about two-thirds of that.

The development of the Web is global. By the end of 1997, there were more than 58 million adults using the Internet in the United States and Canada, while subscriptions to on-line services is on the rise. Japan has also experienced phenomenal growth in the number of Internet users. Europe's on-line revenues from business trade, consumer retail, and content is projected to climb from a combined $1.2 billion in 1998 to $64.4 billion in 2001.

This means that electronic commerce seems to have come of age [1] and consumers are ready to use a new breed of services developed in many cases by the same firms that used to provide them to business firms.

Such a move is not unprecedented. As various authors have remarked, services, not manufactured goods, have fueled modern economic growth [2]. The Industrial Revolution of the eighteenth century involved changes not only in production, but also in financial structures, and in transportation and communication networks.

It is no coincidence that two of the biggest service sectors, banks and railroads, were booming at the same time as the Industrial Revolution. Without the emergence of these and other services, the economic benefits of large-scale production units could never have been realized [3].

One may argue that we are experiencing the same pattern with today's Information Revolution, which involves change not only in information production but also in information uses.

In this matter, America is clearly showing the way. According to *Fortune* magazine's yearly ranking of the 500 biggest American companies, in 1996 the sales figures of the major companies in computer and data services, electronics networks, and telecommunications comprised about 64% of the sales of the major motor vehicle firms. In a similar ranking the same year of the 500 biggest companies globally, the sales of the major high-tech services companies comprised only 45% of the sales of the major motor vehicle firms. This leadership in high-tech services is probably one of the reasons why the United States moved ahead in the 1990s while Japan and Europe were stagnant.

Since computers or network solutions are getting more ubiquitous, the value of information for customers, whether firms or consumers, is being redefined as it moves from simple availability to more sophisticated content.

Furthermore, as stated by one executive from the service division of Compaq: "Today, high-technology products are entering a commodization phase. Everyone has become better at developing products. The one place you can differentiate yourself is in the service you provide." As the reader will discover throughout this book, in the information industry there are numerous examples of high-tech firms that are using services to differentiate themselves from the market and to increase profit.

References

[1] Harrington, L., and G. Reed, "Electronic commerce (finally) comes of age," *The McKinsey Quarterly Review*, No.2, 1996, p. 68.

[2] Bateson, J. E. L., *Managing Services Marketing*, Orlando: The Dryden Press, 1995.

[3] Riddle, D. I., *Service-Led Growth: The Role of the Service Sector in World Development,* New York: Praeger Publication,1986.

1

Understanding the Nature of High-Tech Services

THE HIGH-TECH SERVICES BOOM looks set to continue; it seems likely that there will be no successful business that does not make service the foundation of its competitive strategy. For instance, it is widely accepted that Compaq, the leading PC maker, bought Compaq in 1997 for the astonishing sum of $9.6 billion mainly because of the value of Compaq's service business. Similarly, European computer manufacturers like the French company, Bull, or the German company, Siemens-Nixdorf, have sold their PC division to concentrate on the development of their service activities.

Even software editors are developing their service activities. The service revenues are only 10% of the total revenues of Microsoft but they represent 37% of the total revenues of SAP, the German software company and worldwide leader of Enterprise Resource Planning (ERP), an integrated application software that allows corporate customers to manage all the processes and functions of the firm as a whole.

Finally, a company like AOL has built its entire success in redefining the offer of digital services to nontechnical consumers while IBM is entering the Internet market and electronic commerce with its "e-business" range of services directed to businesses, as well as to end-user consumers.

Other high-tech services are also experiencing growth and success such as vehicle satellite positioning and tracking, advertising delivery, and video backhaul transmission via satellite or fiber optic cable.

Surprisingly, these new types of services are not considered much in the literature about services. Traditionally, those books focus much more on what could be called more "traditional" services like the ones offered in the fast food, restaurant, hotel, travel and leisure, or transportation industries. While high-tech services clearly belong to the service category, it appears they have some specificity worth a detailed and careful study in order to achieve better management of those services.

But before giving a definition of high-tech services, it is important to understand the nature of services.

1.1 What are services?

To define the sum and substance of services, consultants and academics who have studied and analyzed service firms and businesses have focused on the main differences between services and goods.

But, we must note that in actuality it is extremely difficult to define a pure good or a pure service. A pure good implies that the customer obtains benefit from the good alone, without any added value from service; concurrently, a pure service assumes that there is no "goods" element to the service that the customer receives. In reality, most services contain some goods elements, and most goods offer some services—even if it is only delivery—more specifically for high-tech products.

The goods and services dichotomy is a subtly changing spectrum with firms moving their position within that spectrum over time. In fact, an exact definition of services is not really necessary in order to understand them.

That there are different problems associated with the two is readily apparent. Mills and Moberg [1] describe two factors that set services operations apart from goods operations, namely, differences in *output* and differences in *process*.

1.1.1 Services as an output

Indeed, as stated by one leading researcher in the service area, C. Lovelock, "To differentiate services from goods, one can first consider the very nature of the output as *a performance at a given time* rather than an object. In other words, a service is intangible and what is important for the customer is the result more than the process."

The second key characteristic is that the performance of a service is delivered through an interactive *experience* involving the customer to a greater or lesser extent.

These two attributes, being both a performance and an experience, produce consequences that clearly distinguish a service from a good (see Figure 1.1).

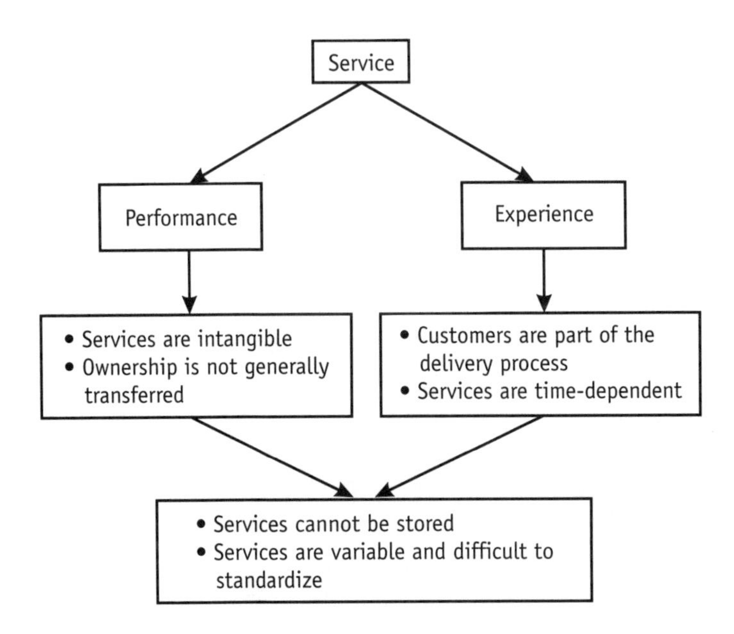

Figure 1.1 The attributes of a service.

1.1.1.1 Services are intangible

In essence, services are invisible since they are a performance. Accordingly, it is not possible to use packaging to communicate the nature of a service. Similarly, services cannot be patented and, consequently, are easy to duplicate.

Actually, as we have seen before, a pure service does not really exist. There are always some material goods attached to a service, like the room for a hotel, or the plane and the seat for an airline transportation service. The service offering, which is what any marketing proposition must consider, is always a mix of services and goods.

But it is important to keep in mind that the material part of a service never encompasses the full reality of the performance of the service. The only physical trace of a concert performance of an orchestra is usually the leaflet with the program (and sometimes, exceptionally, a live recording); it does not give any clue about the quality of the performance on a given day, be it wonderful, average, or atrocious. In services, what you see is never what you get.

1.1.1.2 The ownership of a service is not generally transmitted

A customer never actually owns a service because of its intangibility. This is a major difference with a product whose possession always has a value, whatever its performance, because it can be shown (providing status) or, it can be seen (providing a feeling of beauty, or strength, or confidence, and so on). As a consequence, service users and buyers focus much more on the result, the performance, and the output of the service than on the way it is delivered because that is where the only value is.

1.1.1.3 Customers are part of the delivery process

Services deliver a bundle of benefits to the customer through the experience which is created for that customer. The way in which the customer receives the benefits package is thus very different for services than it is for goods. With goods, the benefits package is intimately connected to the actual goods and remains a part of it, generally disappearing once the good has been consumed or is not being used.

With services, however, the benefits can come from a variety of sources interacting together at once according to the so-called "servuction" service model [2]: the environment, the contact personnel, (the latter being the part of the service "factory" encountered by customers), the other customers, and the back-office organization, as shown in Figure 1.2.

Because customers are part of the service process:

- Any change in the "service factory" will have an impact on customer behavior. Changes made to the visible part of the service may affect his or her decision-making and purchasing process and frequently will demand that the customer change his or her behavior.
- Any change in the benefit concept—that is, a repositioning—means change in the visible part of the factory.
- Everyone who comes into contact with the customer is delivering the service. Contact personnel are part of the customer experience. Unlike inanimate goods, they have feelings and emotions that are apparent to the customer, can affect the service experience positively or negatively, and are extremely difficult to control, monitor, and standardize.

1.1.1.4 Services are time dependent

Services are very often time dependent because customers want to use the service when they need it. Consequently, sometimes there are peak times when too many customers want to access a service at once creating a bottleneck.

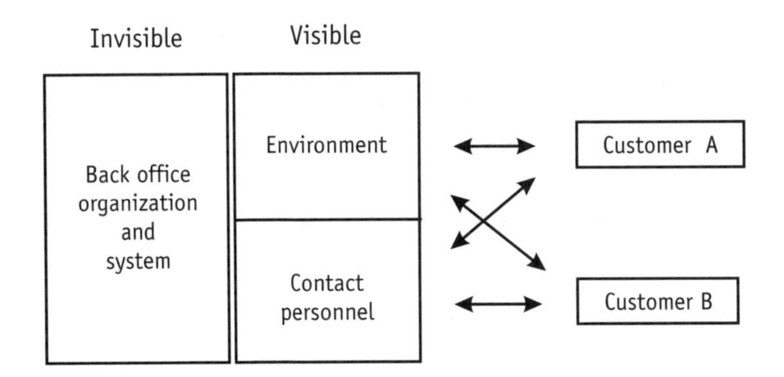

Figure 1.2 The servuction model.

In these cases, generating further demand can be more a disadvantage than an advantage because it puts the resources—either contact people or equipment—under pressure leading to a bad customer experience through delay, or poor quality performance.

1.1.1.5 Services are perishable

In order to receive the benefits of the service, the customer must be part of the service, sometimes not only at the consumption stage but even at the production stage. It is thus impossible to inventory services. This creates difficulties for marketing. In goods, the ability to create an inventory means that production and consumption of the good can be separated in time and space. That is not the case with services where most are consumed at the point of production, though there is an exception for services that can be delivered electronically.

1.1.1.6 Services are variable and difficult to standardize

The intangibility and the absence of inventory also greatly complicate the quality control of services. In a goods-producing organization, the set of minimum variability in the quality of the product and the selection of individual items for testing can be achieved more easily.

Services cannot be fully quality controlled at the factory gate, before reaching the customer. When service occurs in real-time, if something fails during the service process, it is too late to have quality-control measures. Indeed, the customer—or another customer—may be part of the quality problem.

In manufacturing, mistakes tend to be a recurring part of the process, meaning they can be identified and corrected; while in services, errors are isolated and happen randomly. The only answer is to have a perfect process to eradicate mistakes at the source (including customers' mistakes).

The inability to inventory a service makes interaction almost constant for the marketing and production department and, in any case, much more so than in a goods-producing firm. In the latter case, stock is very often sold at a transfer price from one department to another according to a negotiated contract or agreement on the quantity, quality, and delivery schedule.

Once the agreement is reached, each department is able to work relatively independent of the other. This is almost impossible in a service firm, where each department can agree on the availability of some resources at a given time, but must always work on an *ad hoc* (case by case) basis.

But service businesses also differ from each other. All products—both goods and services—consist of a core element that is surrounded by a variety of sometimes optional supplementary elements. But it should be a mistake to examine services only on an industry-by-industry basis because they can be classified according to their type of operational process, a key strategic driver for those activities.

1.1.2 Service as a process: The three categories of services

We can identify three broad categories, depending on the nature of the process and the extent to which customers need to be physically present during service production (see Figure 1.3).

1. *People-processing services* involve tangible actions to customers in person. These services require that customers become part of the production process, which tends to be simultaneous with consumption. In businesses such as passenger transportation, health care, food service, and lodging services, the customer needs to enter the "service factory" (should it be an air terminal, a hospital, a restaurant, or a hotel) and to remain there during service delivery. Either customers must travel to the factory, or, service

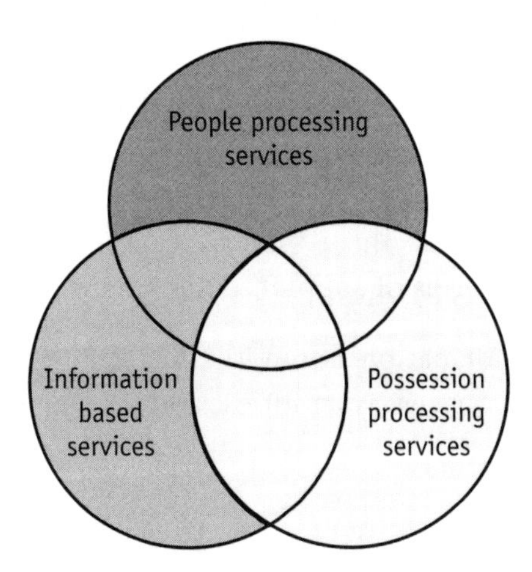

Figure 1.3 A classification of services according to their operational process.

providers and equipment must come to the customer. A representative example of a high-tech people-processing service is information technology consulting or network consulting services where consultants come to customer locations.

2. *Possession-processing services* involve tangible actions to physical objects like freight transport, warehousing, car repair, laundry, and disposal. The object needs to be involved in the production process, but the customer does not since the consumption of the output tends to follow production. Again, the service "factory" may be fixed or mobile. A local geographic presence is required when the supplier needs to provide a service to physical objects in a specific location on a repeated basis. Typical examples of high-tech possession-processing services are computer and telecommunications equipment installation and maintenance.
3. *Information-based services* depend on collecting, manipulating, interpreting, and transmitting data to create value. Examples include accounting, banking, consulting, education, entertainment, insurance, legal services, and news. Customer involvement in production of such services is often minimal. The development of global telecommunications networks makes it possible to use electronic channels to deliver information-based services from a single source to almost any location. Local presence requirement may be limited to a terminal-telephone, fax machine, computer, or more specialized equipment like a bank ATM machine connected to a reliable telecommunications infrastructure with adequate performance.

Complementing these three generic categories of services are the following eight categories of supplementary services [3]:

1. Relevant information about the service, such as the working instructions, prices, warranties, and so on;
2. Order taking;
3. Caring for the customer;
4. Protection of the customer's possessions;

5. Managing exceptions to the normal delivery, should they be special requests or technical problems;
6. Clear and on-time billing;
7. Easy and adapted payment process.

These elements added to the core service contribute to adding value to users while helping successful firms to differentiate themselves from their less sophisticated competitors. These subsidiary services are clearly part of high-tech services.

1.2 What are high-tech services?

In order to define high-tech services, we will use the testimonials of different executives of various service vendors. Let us make it clear up front that we have selected only information technology services companies. This is mostly because we believe that information technologies are the fastest growing categories of high-tech services, both in business and consumer markets. We also believe that most of the lessons learned in those services can be applied to other technology-based services.

We will then consider an extensive list of high-tech services that are currently available on the market. Finally, we will determine the key characteristics of high-tech services as a specific type of service.

1.2.1 High-tech services: A definition

When asked about the definition of high-tech services, most of the executives we have interviewed gave a similar explanation. Consider some of their answers:

- "We are in the business of programs of complex change through complex technology. We are using technology to offer innovative business solutions to our clients. We are using innovative strategic concepts with innovative technology." (Andersen Consulting)
- "High-tech services are services requiring strong individual and collective competence, which is exceptional and expensive, as well as requiring a constant adaptation to innovation." (Steria)

- "We used to define our mission as a vendor of technological value-added services. Recently, we have been acquired by another firm and our mission has evolved to be an information services company helping our clients to use information effectively." (Experian)
- "High-tech services are services implemented by highly technical engineers on innovative hardware and software architecture (i.e., less than 3 years old)." (Syseca)
- "Our high-tech services activity aims at creating and implementing business solutions relying on technology and having a strong and effective impact on the operational activity of our corporate clients." (IBM Global Services)
- "Being in the high-tech services business means understanding customer needs and bringing value through solutions that mix methodology, personnel, and information." (Compaq)

Though the wording may be different, all of the executives of these leading information services firms label their business as *offering value to their customers* through services based on *innovative information technology* (hardware and software) and implemented by *highly knowledgeable personnel* relying heavily on *methodology*.

1.2.2 High-tech services: A list

Before analyzing this new breed of services and considering if they fit with the standard model of analysis of more traditional services, let us consider the range of services currently offered on the market (see Figure 1.4). A clear distinction must be made between services directed to businesses and those directed to consumers who have different needs and expectations.

1.2.2.1 High-tech services for businesses

The main services offered to corporations and organizations are the following:

- *Consulting* services help firms to adapt to changes in their market, their environment, and their own structure. Consulting services

Type of services \ Type of customers	Businesses	Consumers
Strategic consulting	○	
System engineering	○	
System integration	○	
Support	○	○
Outsourcing	○	
Network services:		
• On-line information	○	○
• E-transaction	○	○
• E-business	○	○

Figure 1.4 A list of the main high-tech services offered to customers.

involve business processes design, advice on marketing, change management, training and reskilling, and the use of information technology.

- *Systems engineering services* assist customers in the design and running of information systems. They include capacity planning, system management, and help desk.
- *Systems integration services* help customers to make various information technologies working together to achieve a business solution and performance. Today, information technologies encompass very different elements like computer and communication hardware, operating and application software, and multimedia technologies. Some information technologies can be off-the-shelf solutions, others need to be tailored, while others are purely specific. Mixing all those technological ingredients without the adequate knowledge can turn out to be a nightmare for customers.
- *Support services* are designed to maintain the technological solutions up and running for customers. These services include hardware and software maintenance, restoration, relocation, and even

replacement. They also include all the back-up services and the business recovery services that enable customers to keep on working, even after a major disaster.

- *Outsourcing services* foster the ability for customers to concentrate on their core business. Outsourcing also allows them to hand over all or part of their information systems under long-term contracts. Outsourcing began with mainframe systems and spread to all information technology equipment like networks, servers, or desktop computers. Recently, outsourcing has expanded to include the management of entire business functions and not only the information technology aspects.
- *Network services* let corporate customers enjoy the benefits of network computing. They cover services such as data network management, electronic transactions, workgroup applications, hosting, and Internet connections.
- *E-business services* developed customized solutions based on the Internet for communication (Intranet and Extranet), security, or electronic commerce. E-business services also embody electronic intermediation. Companies like Industry.Net or IBEX are shaping business-to-business markets by electronically facilitating the communications, negotiations, and eventually—when secure process software becomes available—transactions between buyers and sellers. Other companies are offering hosting services for e-commerce such as Open Market, which provides Internet software, servers, and applications allowing firms to outsource their e-commerce activity if they choose.

1.2.2.2 High-tech services for consumers

Information technology services are also offered to consumers of the exploding Web market. The bulk of these services is constituted by:

- *On-line information services* (including publishing and advertising) provided by electronic journals and periodicals, on-line catalogs, or information brokerages (whose value is to identify the various sources of information).

- *Electronic-transactions* where no monetary exchanges are directly involved such as on-line "help desks" (which return maintenance solutions via electronic bulletin boards), on-line customer satisfaction feedback surveys, or job postings.
- *Electronic-business* that is best represented by electronic storefronts where customers can remotely demonstrate, purchase, and immediately download information products that can be transported in a digital format, including text, audio and video. Home banking, investment services, and computer software are good examples of information product. Under the current Internet pricing regime, the marginal cost of delivering them is practically zero.

1.2.3 The attributes of high-tech services

To identify the various attributes of high-tech services, we will use the analytical framework used to describe the services in general. This will allow us to see how high-tech services fit with this model and will help us to single out the peculiarities of these services (see Figure 1.5).

1.2.3.1 High-tech services are intangible

All the executives we have interviewed agree on the fact that basically the services they are offering are very much intangible for the customer because they are very much associated with intellectual property. As stated by an executive from Andersen Consulting: "We are selling an intangible solution, namely, a commitment to performance, materialized only through some binders and a written proposal including budgets, costs, and delays."

However, one must note that some high-tech services are more tangible than others (see Figure 1.5). The mix between goods and products differ significantly according to the type of service considered. For instance, one executive at Steria noted: "Application software programs can be fully integrated with hardware, for instance, as in the airplane braking system software program we have developed for Airbus (the European airplane manufacturer). This program is encoded on an electronic card (board) and is fully associated with the hardware."

A major consequence of the relative virtuality of high-tech services is that customers are always reluctant to buy something they cannot see.

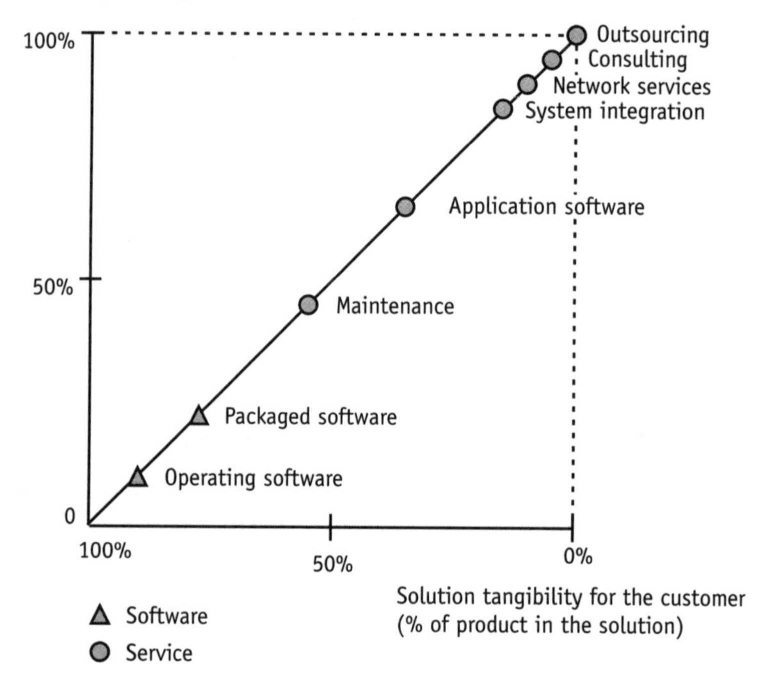

Figure 1.5 The various states of tangibility of high-tech services.

For instance, an executive at Experian said: "A customer is willing to buy a new reporting system which can always be shown, but will never pay for a technological change, considering it to be our problem, not his or hers." Such a trend has significant implications on the billing of the service and leads to some balancing between the visible and the invisible parts of the performance. The invisible part tends to be undercharged and the associated costs are affected to the tangible aspects of the delivery.

Beyond the virtuality of many high-tech services, ultimately what is important is their value to the customer. As one executive of Compaq remarked: "We are currently living in the paradigm that information is intangible because it cannot be touched. For instance, if you cross the border with a 200 megabyte disk, the customs officer will consider only the value of the disk for taxes without taking into account the value of the information on the disk. This value can be one million times superior to the actual value of the disk if it is data decryption software, for instance. Since information has a value, it means that it is intellectually tangible."

1.2.3.2 Ownership is not transferred in high-tech services

Since the customer is buying performance and not know-how, the transfer of ownership is very limited to the tangible output of the service, namely a documentation kit, a written report, or sometimes the code lines of application software. However, all the knowledge encoded in the solution generally belongs to the vendor who keeps the copyrights of the process. The less tangible the service, the less ownership of the service the customer has, for instance, as in the case of outsourcing or network services.

In some very limited cases, the transfer of ownership may happen when a customer requires by contract the exclusivity of the solution developed.

1.2.3.3 Customers are associated with high-tech services

Customers are also part of the high-tech service, most specifically they are very often the ones who must activate some functions and perform various operations on their screen to have the service delivering value.

This is very obvious with Internet services where the user is also the actor: if he or she is not able to correctly perform the various procedures and actions requested on the screen, such as entering a correct e-mail address, giving the right password, or downloading requested software, the service will not function.

But this is also the case with corporate customers. All of the executives of business-to-business high-tech services firms agreed that production and "consumption," or, the use of services, are very closely linked.

Such an intimate connection between the user and the service offered has significant consequences on the management of the customer relationship and the management of the operation.

For instance, "push" technology available on network services is building on this feature and aims to offer fine-tuned data exactly adapted to each user. Push technology is a type of software that enables the content provider to push their content in some way to the user. The user selects what types of content and which content providers they wish to receive content from, and the provider delivers it to the user's computer via the push software.

A common example of this on the Web is the PointCast program, which is similar to a screensaver from CNN. When the screensaver is activated, current headlines, stock quotes, and sports headlines are displayed along with advertisements. Another application of push technology is system administration: push allows for software updates, operating system configuration, and other similar tasks to be automated in software.

1.2.3.4 High-tech services are location independent but time dependent

Electronics allow services to be produced in one locality to be consumed in another, which makes their management easier but those services are still very much time dependent.

Services are very often time dependent because customers want to use the service when they need it. Consequently, there are occasionally rush times when too many customers want to access a service simultaneously, which creates a bottleneck.

This is typical of the telephone business, the Internet, telecommunication networks, and other "big" systems that deliver services. Consequently, it is extremely important to match the level of resources with the expected demand. This ability is generally achieved through many (and sometimes painful) experiences. As an oversupply of resources or customers can be extremely costly, such a matching capacity is very often a competitive advantage for high-tech services suppliers which are ahead of the experience curve.

The case of a telecommunications network service whose demand fluctuates a long time is illustrated in Figure 1.6. If the network capacity stays as a maximum to the position C, the level is short of meeting customers' demands and the network is oversaturated (which may cause it to break down). On the contrary, if the capacity reaches level A, the network is underused and costly since fixed volume is not used often enough. The right capacity is around the B level.

An executive of Atos said: "For calculating the profitability, there is a lot of anteriority in the measures of consumption of our services." Such is one of the benefits of being in the business since a significant amount of time facilitates the use of the right metrics to anticipate the needs in infrastructure capacity for a given high-tech service (see Figure 1.6).

1.2.3.5 High-tech services are relatively homogeneous so that they can be stored and quality controlled

In striking contrast with the general vision of services, many executives of high-tech services firms consider that the services they are offering are not heterogeneous. Consequently they are not perishable and can even be stored because they are putting a lot of emphasis on the methodology of service development and implementation.

One executive of Steria mentioned: "Our services are not heterogeneous because we are delivering solutions based on quality plans, functional analysis, tests which are getting increasingly standardized, even automated, like for the coding of software programs. Of course, we do have some contact personnel at customers' locations, but they are working according to strict rules and methods. This *'intuitu personae'* aspect does not represent more than 20% of a job vs. 80% of controlled production."

Furthermore, high-tech services firms try to industrialize their know-how through the building of knowledge assets and the creation of solution centers based on project analysis. Thus, they manage to increase

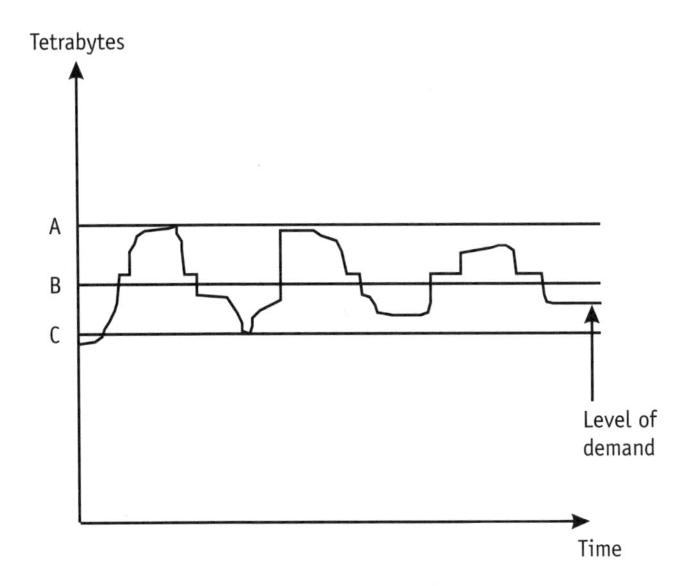

Figure 1.6 The impact of time on high-tech services.

the re-use of the same elements of the solutions they have developed for other customers like application codes, project management tools, implementation procedures, or even the management of human resources with problems in connection with outsourcing contracts.

As one executive of Andersen Consulting said: "In this business, it is a little bit like playing with Lego bricks. You try to define some building blocks and then you combine them differently to achieve various customized solutions."

Besides, as an executive of IBM Global Services noted: "Some of our solutions are already close to products. They are made up of operations which are described and almost constant in time, such as the Year 2000 migration programs, The Euro-currency adaptation program, business intelligence tools like data-mining software, or the ERP programs." However, this executive is quick to remark that high-tech services can not be fully standardized because they must be tailored to the various needs of large customers who have very different businesses and environments and very often require large project solutions.

1.2.3.6 High-tech services cannot be easily demonstrated before purchasing

Because high-tech services are both a performance and an experience, they cannot be easily demonstrated before purchasing, at least when they are customized as is the case for many services sold to corporate customers.

This fact drives two important consequences. The first one is about the selling of those high-tech services. As one executive stated: "In our business, the selling is relatively easy, it is the delivery which is difficult." Consequently, in many firms like Andersen Consulting, for example, the team that sells the solution is the one that also delivers it. It is not possible to separate the selling from the delivery such as in the computer hardware industry for instance, because it would be tantalizing to sell a promise that could not be actually delivered [4]. Such a way could be dangerous for the vendor both financially, if the project goes over budget or the customer asks for compensation, and commercially, because it endangers the credibility of the firm.

This leads us to the second consequence, which is that in the high-tech services business, the image of the vendor is a key element of success.

Because the performance of a proposed service can rarely be tested in advance, a customer will rely heavily on the reputation of the firm, that is, its ability to keep up its commitment to deliver a quality solution on time. And as another executive mentioned: "It take years to build a good image, but you can lose everything on one project alone." The importance of the image as a way to reassure the customer is intensified by the fact that high-tech services include a good deal of innovative technology. But it has been shown that technology, should it be in goods or services, has a tendency to worry customers, especially nontechnical and risk averse ones, and that the confidence in the vendor is a key purchasing criteria [5]. Such is also the case with consumers' high-tech services as we will see in Chapter 2.

1.2.4 The strategic trade-off in high-tech services

Many executives consider that the theoretical line between information-based and people-processing services does not apply when it comes to defining high-tech services. As one executive of Syseca observed: "There is a lot of room for variation between pure data processing and the leasing of information systems specialists. Furthermore, such a differentiation has been blurred by the decrease in the use of mainframes and the outgrowth of client-server distributed architectures. We are offering both information-based and people-processing services as well as system integration services which are a mix of the two."

More relevant differentiation criteria seem to be the weighted asset requirements. Some heavily automated services require low assets, mostly skilled engineers and technicians, as in the case of the credit card management systems, or as in the hosting and running of Internet servers.

Other services are capital intensive such as in facility management or back-up services that require buildings, equipment, and a significant number of operators, who can be located in a front office as in the case of a call center, or in a back office, where the service involves the manipulation of documents like checks or telephone bills. A company such as Experian has more than 1,000 operators running its call center operations, as well as more than 500 people for "processing" paychecks.

The tradeoff of assets vs. skilled engineers and technicians is a vital strategic issue for top managers at high-tech services vendors. Indeed, it defines the mission of the firm, as well as its required level and allocation of resources in competence and in cash. It also defines the scope of the competitive fields the firm will enter. So far, two dominant models of vendors have emerged: specialists and one-stop shopping.

Specialist vendors are usually offering only automated services, such as system integration, networking management, or electronic commerce services. They address niche or narrowly segmented markets where they may achieve a significant competitive position with limited assets. They can offer fine-tuned solutions for specific customers or businesses. In order to make a profit, they have to enlarge their portfolio of customers interested in their distinctive, but limited, offer of services.

One-stop shopping vendors include vendors offering a full array of services covering all the business needs of large organizations. This is the case of all the major services suppliers, such as IBM Global Services, Andersen Consulting, and EDS. Clearly, the number of players of that size is limited, making it easier to compete. On the other hand, the offering of an extensive portfolio of solutions implies massive investments not only in technology, but also in people and other kinds of assets. This means the return on investment may be harder to achieve and is usually made over a longer period of time because of the sheer size of the investments. Consequently, one-stop shopping vendors must achieve a long-term relationship with their customers in order to maximize their return. If not, they may not be able to cover their up-front investment.

1.3 The new rules of high-tech services

It is important to understand that high-tech services and most specifically information-based services obey different rules than product-based solutions. Indeed, because they have a specific way to create value for the customer, along what can be called a virtual value chain, they have slightly different economics.

1.3.1 The virtual value chain model

The value delivered to customers by high-tech services is created with information. But information is a supporting element in the value adding process, not as a source of value in itself for the customer. Simply put, the job of high-tech services vendors is to turn raw information into services for their customers. In the case of corporate customers, the service providers are building a virtual value chain model, which mirrors the physical value chain of their clients.

The value chain analysis helps to describe the various activities within a firm, and to assess their performance when combined into a system to produce value for money solutions.

According to the now traditional model introduced by M. Porter [6], there are five categories of *primary activities:*

1. *Inbound logistics* receive, store, and distribute the inputs which are then transformed by
2. *Operations* into the final product or service through manufacturing, assembly, packaging, and so on.
3. *Outbound logistics* store and physically distribute the solution to the customer while
4. *Marketing and sales* make customers aware of the solution and provide them with the way to buy it and
5. *Services* maintain or increase the value of the solution through installation, maintenance, or training.

Each of these essential activities are linked to support activities of four different kinds:

1. Procurement whose mission is to acquire all the primary resources according to processes like purchasing;
2. Technology development which may concern either product development or process development;
3. Human resources management to recruit, manage, and develop firm personnel;
4. Infrastructure which sustains the organization and the firm culture, including departments like accounting and finance, legal, quality control, or information management systems.

This model helps firms to pinpoint key activities and their interrelations with others to bring maximum value to their customers in comparison with competitors. It allows them to identify the core competencies required to perform in a given business, which may have to be made or bought.

To create value with information, high-tech services vendors are creating a virtual value chain [7] mirroring the physical one of their clients (see Figure 1.7).

First, they propose large-scale information systems, which allow their corporate client to get a better view and coordination of the material operations in the physical value chain. A good example, now also available for consumers, is the electronic tracking of packages or materials from one place to another all over the world and in real time.

In a second step, their service offer is to substitute virtual activities for physical ones, thus creating a parallel value chain in the "marketspace." For instance, through highly powerful and sophisticated CAD/CAM global network services, large corporations are able to develop new prototypes for airplanes, cars, satellites, or molecules on computers, rather than with physical prototypes. They have moved one key element of the physical value chain—product development—into the virtual value chain of the world of information, sometimes called the "marketspace," in opposition to the physical "marketplace."

Rather than national product teams, a virtual team made up of various people located in different places are communicating and working

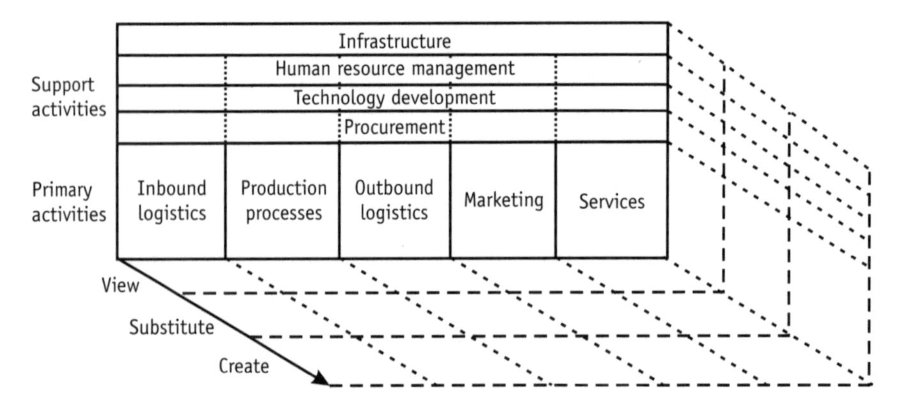

Figure 1.7 From the physical to the virtual value chain.

together to develop new products or services. Developers in a virtual team do not experience the limitations of time and space that characterize management in the physical world. They build and test prototypes, share design and data with colleagues around the world over a computer network 24 hours a day, and receive customer feedback from the other side of the world.

All elements can be moved in the virtual value chain. Thanks to high-tech services, animation movies are made the same way with various teams working together in different places of the world. Global marketing campaigns are conceived, designed, and implemented the same way allowing the same message to be sent all over the world the same day, through video conferencing services for instance. Many maintenance operations on industrial equipment are performed automatically through network services from one remote control location.

Support activities can also be made virtual: global companies are interviewing new applicants all over the world through video-conferencing, while finance, legal and quality departments rely heavily on electronic services to monitor closely and repeatedly the performance of the organization.

The last step is to extract information of one stage of the virtual value chain and turn it into new spin-off products or services. For example, digitally captured product designs can be converted or adapted as multimedia software for PC or video game stations. Similarly, parcel automated tracking data used originally for internal logistics can be repackaged and made available (or sold) on line to consumers. To get the maximum yield, information must be processed systematically. High-tech services help corporate users to:

- Collect information;
- Choose information according to its value;
- Compose information for the users;
- Consolidate information in a usable package;
- Communicate or deliver information to users or potential customers.

1.3.2 The new economics of high-tech services

Relying heavily on information of high-tech services drives some specific consequences because of the so-called law of digital assets. Mostly those services permit massive economies of scale and scope, new dynamics, and increasing returns.

1.3.2.1 The law of digital assets

Simply put, this law declares that digital assets, unlike physical ones, are not used in their consumption. Companies that create value with information can duplicate it at almost nil variable cost and thus offer it through an almost infinite number of transactions. They can even provide the information for free in order to boost the acceptance curve of a new information product or service like Netscape did.

In the world of information, volumes are not the limitations, property and copyrights are. A good example is Amazon.com, the first virtual and "the world's largest bookstore," which has achieved this position with a completely open Web site, not copyrighted, where users can get book summaries and reviews, and can share ideas, as well. However, the books themselves are copyrighted and delivered in a controlled manner.

1.3.2.2 Massive economies of scale

One of the consequences of the law of the digital assets is that information-based services allow radical economies of scale because they can be replicated easily and at almost no cost. In the 1960s it cost about $1 to keep information about an individual customer. Today, it costs less than one cent per customer. Those costs continue to decline sharply as the processing capacity per unit of cost or microprocessors doubles every two years. The substantial decrease in hardware cost is paralleled by a similar trend in software development. According to the executive director in charge at Innovation & New Technologies at Cap Gemini, productivity in software development has risen by 80% every five years over the past 50-year period. Consequently, small companies may achieve low unit cost for products and services in markets dominated by big companies.

Furthermore, network services, such as the Web or proprietary networks offer an extraordinary opportunity for suppliers to gain direct access to consumers, without the attendant costs associated with the maintenance of physical distribution channels. The greatest strength of electronic business or transaction services is the location-independence of the provider. A single virtual shop-front operating 24 hours a day can easily be set up on the Web. This, in turn, avoids fixed costs of staffing and rental expenses associated with a traditional retail format and also generates huge economies whatever the size of the vendor.

Outsourcing services also generate massive return on investments for customers through savings in time, people, and technology resources. According to IBM Global Services, one of its customers, a U.S. steel company, went from the least profitable firm in its industry to the most profitable. Another customer was working with 156 different hardware vendors, 89 software packages (and thousands of releases), and six incompatible e-mail packages before it moved to IBM Global Services as the single point for software, help desk, network management, maintenance, move/add/changes, and consulting.

High-tech services also generate new economies of scope because they are not limited by geography. One service can be made available worldwide through an electronic medium. But people resources can also be spread all over the world: a virtual team of employees based in different locations can work together for global customers in different locations. The flip side for high-tech services vendors is that competitors can emerge from anywhere in the world.

1.3.2.3 The law of increasing returns

Because of their nature, high-tech services are very often in the position to produce increasing returns on investments by quickly achieving a critical mass, which will have a snowball effect. As more people sample and select a service, it generates interest, excitement, and an installed base that attracts still more people—unleashing positive feedback loops that increase momentum. Figure 1.8 illustrates the typical example of an electronic transaction service for instance. In this case, a new event like the extension of international coverage through the network stimulates interest for the service by providing new opportunities for contact with other people from different countries and locations.

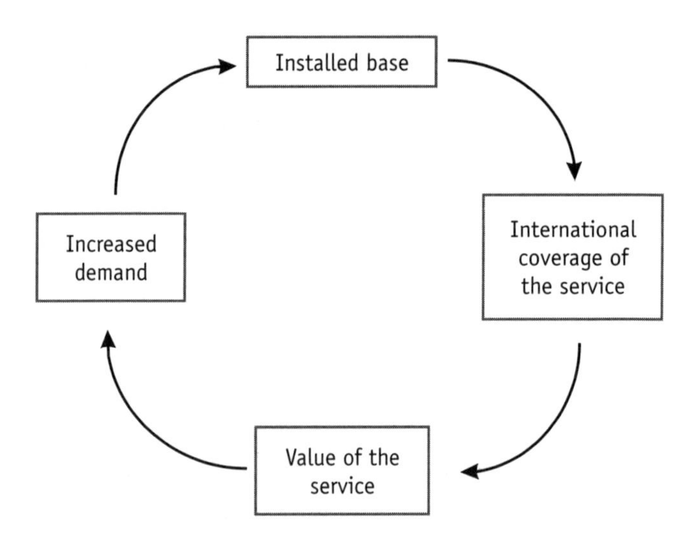

Figure 1.8 Increasing returns for electronic transaction services.

This raises the value of the service for its users but also for prospects, who may be attracted by the service just for its extended coverage. This increased demand develops the installed base, which may become more and more international, and positive feedback has been started which will translate into revenues and profit (provided the vendor has enough capacity to accommodate the growing number of customers).

And when services get distributed over the Internet or any telecommunication network, these feedback loops operate with remarkable speed. Think of how Netscape's engineers could begin writing code in April 1994, introduce their first product in December, and hold a 75% market share by April 1995 before Microsoft retaliates with the same strategy (see Figure 1.8).

1.3.2.4 New dynamics through interactions

High-tech services are no stand-alone solutions, specifically on the Internet. As more services and devices interconnect and depend on one another, they develop interactions that no one can anticipate—and that becomes the basis for entirely new applications. The Internet should evolve like the telephone: two telephones enable a conversation; millions of telephones become the basis for a vast array of services that revolutionize how people communicate. More of something inevitably turns it into something different.

Within large organizations or through the Internet, high-tech services enable greater participation among larger and larger groups of people. That's why millions of people are getting into the habit of visiting the World Wide Web every day. The Web transforms itself, but the changes are co-constructed by the users. No one is in charge and almost anyone can find something of value.

A similar trend is clearly at work when high-tech services are implemented. That's why the intranet, messaging and workgroup services, have become the fastest growing market in new services. They allow individuals to shape and share information and ideas more quickly than ever before—a principle that winning organizations embrace rather than resist.

1.4 Conclusion: The key success factors in high-tech services

We have given a definition of high-tech services and we have described their different features. It is time now to consider them from a management perspective in order to determine the conditions for elaborating this new breed of services successfully.

When we asked them to rank the key winning factors in their business (see Figure 1.10), all the executives of the various high-tech services vendors put customer orientation in the first place. It does not really come as a surprise since they are in the service business, which means that they are selling a performance, not a product, to their customers. However, in a business which is driven by the mastering of state-of-the-art technology, one could have imagined that the technical quality of the service would have come first (see Figure 1.9).

Actually, it comes second before the management of the employees as the third condition of success. Such a ranking illustrates the importance of technology for this category of services compared with traditional services where people management is very often more important. However, some high-tech services—such as consulting, system integration, or maintenance—rely heavily on personnel.

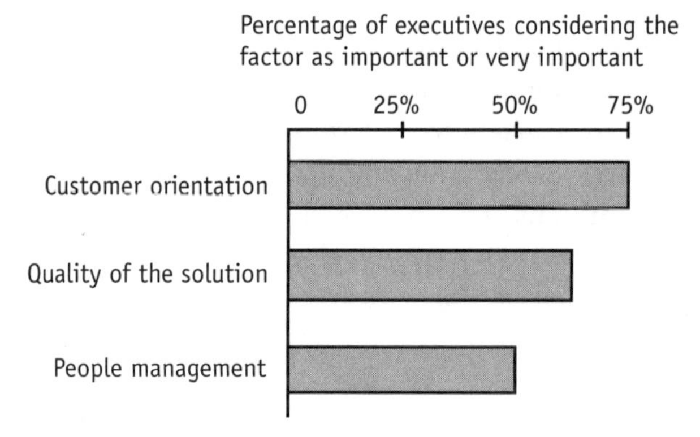

Figure 1.9 The key winning factors in the high-tech services business.

Consequently, we will use this ranking as the architecture of this book. Chapter 2 details how to place customers at the center of high-tech services. Chapter 3 examines the different ways to build quality for those services. Chapter 4 explores the best manner to organize the human resources, while Chapter 5 describes how to effectively manage those resources. Finally, Chapter 6 will consider the peculiarities of marketing high-tech services.

1.5 Chapter summary

High-tech services are experiencing incredible growth and success. However, these new types of services are not considered much in the literature about services. This chapter summarizes the key features of services before it provides the reader with a definition of high-tech services. Then it concentrates on the specific business rules for those new services and concludes by pinpointing the key winning factors needed to master high-tech services.

It is extremely difficult to define a pure good or a pure service. A pure service assumes that there is no "goods" element to the service, which the customer receives. In reality, most services contain some goods elements. Two factors that set service operations apart from goods operations, are namely, differences in output and differences in process.

As an output, services can be defined as a performance at a given time, rather than an object, through an interactive experience involving the customer to a greater or lesser extent. These two attributes, being both a performance and an experience, drive a certain amount of consequences that clearly distinguish a service from a good:

- Services are intangible.
- The ownership of a service is not generally transmitted.
- Customers are part of the delivery process.
- Services are time dependent.
- Services cannot be stored.
- Services are variable and difficult to standardize.

As a process, three broad categories of services may be identified:

1. People-processing services involve tangible actions to customers in person.
2. Possession-processing services involve tangible actions to physical objects.
3. Information-based services depend on collecting, manipulating, interpreting, and transmitting data to create value.

Complementing those three generic categories of services are some supplementary services.

High-tech services can be defined more precisely as *offering value to a customer* through *innovative information technology* (hardware and software) implemented by *highly knowledgeable personnel* relying heavily on *methodology*.

A list of the most significant high-tech services includes professional services such as consulting, systems engineering, systems integration, support, outsourcing, network, and electronic-business services. Consumer services are mostly on-line information, electronic transactions, and electronic business services.

- High-tech services share some of the following features of services:
- They are intangible.
- Ownership is not transferred.
- Customers are associated with them.
- They are location independent and time dependent.
- They are relatively homogeneous so they can be stored and quality controlled.
- They cannot be easily demonstrated before purchasing.

The trade-off of assets vs. skilled engineers and technicians is a vital strategic issue for top managers at high-tech services vendors. Indeed, it defines the mission of the firm, as well as its required level and allocation of resources in competence and in cash. It also defines the scope of the competitive field the firm will enter. So far, two dominant models of vendors have emerged: specialists and one-stop shopping.

High-tech services, and most specifically information-based services, obey different rules than physical solutions. First, they have a specific way to build value for the customer, turning raw information into services by creating a virtual value chain for corporate customers. Secondly, high-tech services have slightly different economics than traditional services because they are ruled by the law of digital assets, which states that digital assets, unlike physical ones, are not used in their consumption. This fact translates into massive economies of scale, increasing returns, and new market dynamics through interactions.

Finally, the executives of the high-tech services vendors rank customer orientation, technical quality, and people management as the three winning factors in their business.

References

[1] Mills, P. K., and D. J. Moberg, "Perspectives on the Technology of Service Operations," *Academy of Management Review,* Vol. 7, No. 3, 1982, pp. 467–478.

[2] Langeard, E., and P. Eiglier, "*Servuction,*" Paris: McGraw-Hill, 1987.

[3] Lovelock, C., and G. S. Yip, "Developing Global Strategies for Service Businesses," *California Management Review*, Vol. 38, No. 2, Winter 1996.

[4] O'Shea, J., and C. Madigan, *Dangerous Company*, New York: Times Business, 1997.

[5] Viardot, E., *Successful Marketing Strategy for High-Tech Firms*, Norwood, MA: Artech House, 1998.

[6] Porter, M., *Competitive advantage,* New-York: The Free Press, 1985.

[7] Rapport, J. F., and J. J. Sviokla, "The Virtual Value Chain," *Harvard Business Review,* November–December 1995.

2

Placing Customers at the Center of High-Tech Services

ANY SUCCESSFUL HIGH-TECH SERVICES COMPANY has figured out that, as stated by the CEO of Andersen Consulting in its annual report, "if we help our clients be successful, then we'll be successful in return." Conversely, a poor service performance may result in a customer's business failure.

In any business, the key to strategic success is to bring value to the customer. Value is what the customer gets for his or her money, (i.e., the benefits provided minus the cost of acquisition). This value will be different for each customer, according to their expectations and personal experience. Consequently, it is critical for high-tech services firms to understand the perceived value of the service by each customer or categories of consumers, to determine the right offer price and the attached margin which will lead to a healthy profit.

The perceived value of a service will fluctuate according to four different influences. The first category is related to the customer's expectations about the way a service should be delivered by a particular firm or industry. Secondly, the comparison with the service offers of competitors may also affect the perception of a service. Thirdly, the vendor's reputation has an impact on the perception of a service, mostly because it is difficult to appraise a service before using it. Ultimately the service performance will set the perception of the value for the customer.

But it is important to understand that in the service business this value is not brought by a product, since the customer is mostly buying a performance, an intangible result. Consequently, the key to success in the high-tech services business does not lie in an aggressive management of the product life cycle like is the case with high-tech products [1], but in an effective management of the customer relationship.

Should it concern consumers or business customers, the very nature of the customer relationship stays the same and goes according to a cycle (described in Section 2.1) which drives some key consequences for the high-tech services firms. However, the implication of those consequences will vary significantly according to the type of customer, individual or organization.

2.1 The customer relationship life cycle

Like for any other services, one can identify three different steps where the customer's behavior will vary significantly towards a high-tech service: prepurchase, consumption, and postpurchase (see Figure 2.1).

The prepurchase step starts when the customer, as a person or an organization, realizes the existence of a need to be fulfilled or a problem to be solved, implying a potential purchase externally. This leads the customer to search for information, identify a collection of solutions, and pick the solution regarded as being the most adequate.

In business-to business activity, such a process is often formalized into a Request for Proposal (RFP) where the customer describes the problems it wants solved and the type of criteria it will consider in order to make the selection among the firms which will bid for offering a solution.

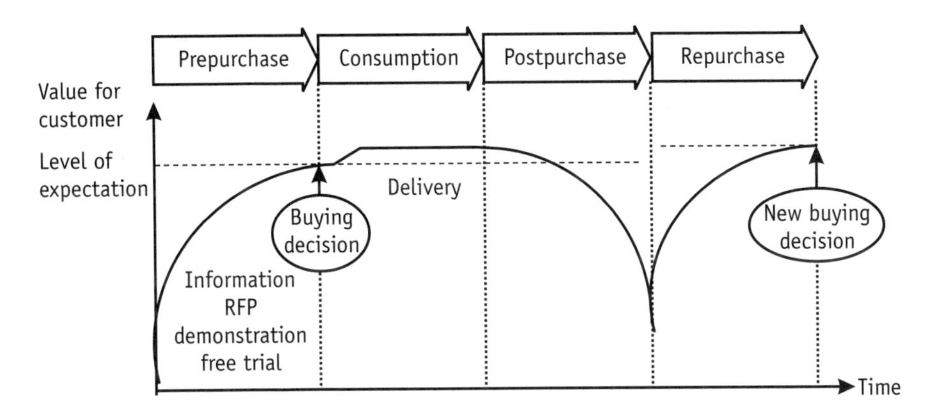

Figure 2.1 The customer relationship life cycle.

Another approach by the customer is to ask for a demonstration or a free trial of a new service in order to make its evaluation.

In the consumption phase, the customer is experiencing the reality of the service it has bought. Like for a product, the customer will compare the service it is receiving with its expectations. Or more precisely, it will compare its perception of the service it is receiving with its expectations. It is important to understand that which matters is not the reality but the perception of the solution by the customer.

For instance, the more positive the perception and the bigger the discrepancy between expectations and cognition, the bigger the satisfaction for the consumer and therefore the perceived value of the service offered. Conversely, a negative perception from a service where a customer had great expectation will translate in very low satisfaction for that customer (see Figure 2.2).

There lies a significant difference within the product business: because a service is a performance created by the interaction of the service firm with the customer, the service firm can influence that evaluation directly during the consumption stage. This is not really possible with a product whose use is generally free of any direct-vendor influence. Actually, it is possible to achieve a similar result by wrapping a service with the product (e.g., Microsoft's Web page or GM's OnStar).

One can differentiate the various type of expectations in three broad categories:

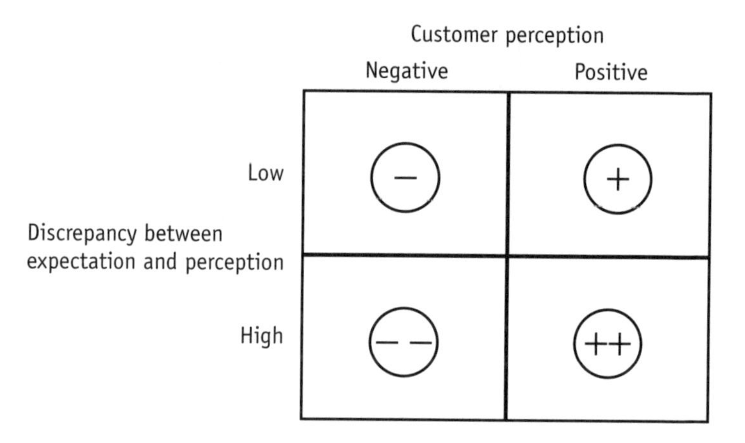

Figure 2.2 The drivers of customer satisfaction and value.

1. *Desired expectations* are what customers assume will happen. They are updated after each use of the service, and are impacted by the competition's offers as well as the evolution of the environment and what is going on in other industries.
2. *Predicted service expectations* are what customers believe ought to happen, according to a normative standard. They are more enduring and vary only when they have been exceeded during one (or many) service experience. An outstanding experience can lead to creating higher predicted expectations.
3. Conversely, *acceptable service expectations* occur when customers figure out that their desire cannot be fully fulfilled and adapted to a lower level of expectation.

The postpurchase stage steps in after the service has been used. In that phase, the customer will assess the quality of the experience it has been through and will make the decision either to use the same service from the same vendor next time it will need it, or to change the vendor or even the service. High-tech services provide information whose time span and value is short and need to be updated frequently. The postpurchase phase is critical to ensure repeat business for service providers. It is at this stage that they will have to develop a loyalty program to keep their customers with them.

To achieve success, a service vendor must manage different priorities in each of the three phases of the customer relationship life cycle. First, the vendor must have an extensive understanding of its customers, not only their identification but also their needs and their decision-making process in order to encourage them to select its service offer.

Once the decision has been made by the customer, the high-tech services vendor must guarantee that the customer will get a positive experience in line with its expectations. If something wrong occurs during the service, the vendor must be able to identify and clear the problem on the spot; this is known as "service recovery" and, if done right by the vendor, can potentially cover a host of sins as we will see later on.

Finally, in the postpurchase phase, the vendor must get a fair assessment of the customer's satisfaction in order to methodically monitor its performance. Then it re-enters in a new prepurchase phase. Let us consider in detail the implications of those three steps.

2.2 Identifying and understanding customers

When considering to gain a better comprehension of its customers, a high-tech services vendor has to differentiate individual consumers from businesses and organizations. Indeed, some characteristics to consider are similar, notably regarding the purchasing behavior of a person, should it be on its own or in an organization. However the *modus operandi*, the buying process is very different in each case and calls for a clear analytical differentiation.

2.2.1 Understanding the high-tech services consumer

When someone buys a high-tech service for personal use, it is influenced by four classes of factors: sociocultural, psychosocial, personal, and psychological (see Figure 2.3). This will be illustrated by an example using on-line services in the United States, the most successful and outstanding category of high-tech services for consumers in the last five years.

On-line services are reaching mass market proportions in the United States. According to a survey of the United States' population by IntelliQuest Information Group, 56 million adults, or 27% of the

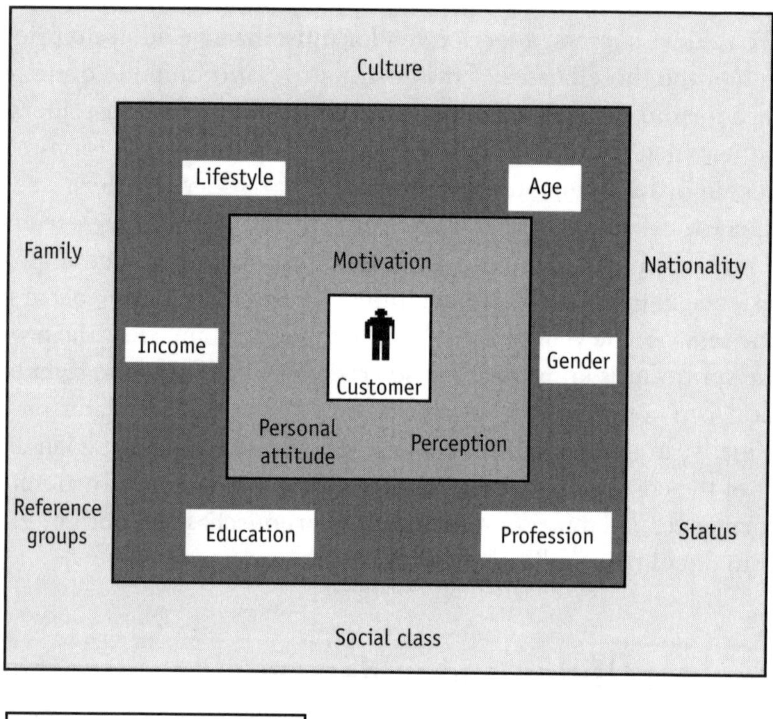

Figure 2.3 Purchasing criteria groups for high-tech consumer products.

United States' population, aged 16 and older are on line using the Internet and on-line services, as of the third quarter of 1997. Another firm, IDC/LINK projects that approximately 40 million U.S. households (or 38% of all households) will subscribe to at least one on-line service by 2001. The average amount of time a user spent on line has grown from an average of 6.9 hours per week in the second quarter of 1996 to a mean level of 9.8 hours per week in the second quarter of 1997.

2.2.1.1 Sociocultural factors

It is true that a lot of technology was requested to make on-line services available to the mass market of consumers. The development of the World Wide Web, cheap 28.8–56K modems, improved Internet backbones, advanced server applications to facilitate on-line services, particularly e-commerce, and, of course, cheap and powerful personal computers.

However, such technical wizardry does not explain why on-line services are so popular today in America and why they are so appealing in other parts of the world. It can be argued that one of the reasons is because our culture values images, should they come from a TV set or a computer screen.

The consumption choices also vary according to nationality, religion, and race. Despite the fact that most content on the Internet is written in English, in December 1997, about 43% of the current 80 million Internet users were from outside the United States, according to Dataquest.

In Europe, Germany remains the leader for Internet access at home, with France still trailing the field. The data on most recent PC purchases reveals that 22% in Germany have access to the Internet or an on-line service compared with 17% in the United Kingdom and only 12% in France.

The intention of existing PC owners to acquire access is, however, high in all countries. Users in regions such as Pacific Asia and Latin America are overcoming some significant infrastructure and regulatory barriers to get on line.

Reference groups (family, neighbors, friends, colleagues, etc.) have a strong influence on the use of high-tech services. Using on-line services can be influenced by family pressure, impressionable neighbors, and friends or colleagues who have already bought one and are very happy with it.

As a matter of fact, in the United States, home is the most popular access location, with 66% of users accessing on-line services from home. The number of households in Britain with access to the Internet has more than doubled from just under 400,000 in June 1996 to 960,000 in June 1997, according to the Britain based NOP Research Group. In France, 570,000 households had access to the Internet in May 1998, representing 2.4% of all households and 11% of PC equipped households, according to Mediangles.

Peer pressure at work also exerts a strong influence through imitation or emulation. While the majority of American users access the Internet from home, the population of users accessing it from work is large and growing fast. For example, 23.3 million people (46% of the on-line population) were going on line while at work, a 57% increase from the same period in 1996. In France, the 1.2 million on-line service users at work outnumbered the 570,000 users at home (not taking into account the users of Minitel, a pioneer technology in supporting on-line services, which is now largely outdated and swiftly replaced by the Internet and private network on-line services).

The *social environment* also plays an important role: someone who belongs to a fairly high class spends more money on leisure activities and is a prime target for a new on-line services.

Furthermore, some on-line services can be perceived as a status symbol which appeals to consumers who buy services for enhancing their social status.

2.2.1.2 Personal factors

Age is an important determining factor. On-line services are of main interest to age groups which invest heavily in leisure activities. A significant majority of on-line services users are young. In France, 56% of Internet users are aged below 40, according to InternetTrak.

Gender is also a key differentiator in on-line services consumption. In the United States, females now account for 47% of Internet and on-line service users, compared with 36% a year ago. This trend should continue, as females make up 58% of nonusers who intend to go on line in the future. However, the electronic market is dominated by males (62% of shoppers and 70% of purchasers) at the end of 1996 since men are still more likely to have access than women.

In the United Kingdom, the number of women as a proportion of the overall Internet user base is also continuing to increase, representing approximately two in five of all current users. In Germany, only 27% of those who have access to the Internet are female, a percentage which is similar to France, where 72% of users are male.

The consumer's *financial status* (level of income and debt) is also important in the decision to purchase a high-tech service. For sure, on-line services are not very costly for themselves. Typically, in the United States, the median monthly expenditure of the on-line user is $50.

But this does not include the cost of membership as well as the cost of the hardware needed to access the service.

Lifestyle also determines the consumption choices. In the beginning of the 1990s, on-line services used to attract more often *"forerunners,"* interested in new technologies and risk-takers than the *"traditionalists"* (more careful and conservative) as far as their buying habits are concerned.

There is a relatively small proportion of extremely active on-line service users (20%) who spend 10 hours or more per week on line, but nearly 40% of all users are spending more time on line than they were last year. A typical AOL user spends an average of 51 minutes per day (up from 16 minutes in 1996). Most of the users find more time to be on line by watching less television, which is a significant shift from the mainstream American lifestyle.

Education is an important purchasing factor as well, because a lot of high-tech services are concerned in dealing with information. As a rule, the level of education of high-tech service users is higher than the average. In France, for instance, 43% of on-line users have a college degree.

On-line services also attract professionals. In the United States, the business/professional segment of on-line consumers registered strong growth in 1997, jumping 21.8% to 3.6 million subscribers, up from 2.9 million in 1996.

2.2.1.3 Psychological factors

The purchasing of a high-tech service may also be influenced by the customers' psychological factors. Important factors to be considered are the motivation, the attitude and the perception of the user or potential user.

Motivations are the reasons why the customer decides to use a service. They are various typologies of needs. The most useful and used is the one set by Maslow who developed a needs analysis grid which is divided into five categories of needs: physiological, safety, love, esteem, and self-actualization. The use of on-line services can therefore respond to many motives. These motives correspond to diverse needs such as reassurance and belonging to a group (by imitating people who already use on-line services), being respected (by differentiating from others who cannot afford a computer necessary to do e-shopping) or treating oneself (by playing a game or listening to music on line).

As a matter of fact, users perform a variety of personal and work-related activities on line no matter where they access from, with the most popular activities including sending/receiving e-mail, obtaining information about hobbies, researching products or services, and accessing general news. In other words, people are doing on line what they are doing off line: communicating, getting information, shopping, and socializing.

The purchase of a service also depends on the *perception* that people have of the service. Freudian theory emphasizes the psychological dimensions of a service, whatever its kind. Accordingly, a buyer of on-line services will take into consideration more than just the performance. The buyer will also consider all the elements capable of activating emotions which will reinforce its attraction for a service or, on the contrary, will keep him from using it. Its perception will be influenced by the design of each screen, the quality of the picture showing a product, and the sequencing of the operation on the Web site.

Herzberg differentiates between the two states of satisfaction and dissatisfaction which exist in each person. The practical consequence is that a company must absolutely avoid dissatisfying elements and must carefully list the satisfying elements for the consumer so that these elements can be added to the product. Therefore, a foreign on-line services set that cannot be understood because of a language problem will bring about dissatisfaction. On the other hand, an extraordinarily attractive picture can provoke the consumer's satisfaction and enthusiasm and will lead him or her to purchase the product on line.

Perception is complicated by two phenomena: *selective distortion* and *selective retention*. Selective distortion makes someone "adjust" information in order for it to correspond to their wants. In Europe, someone who likes the Internet will have a tendency to idealize the advantages and reduce the disadvantages when considering on-line services.

Selective retention leads the customer to better remember information that reinforces their beliefs. An advocate of on-line services will more easily remember the advantages of such a service and the disadvantages of going to the supermarket and vice versa. As we will see later, selective distortion and retention apply to any kind of high-tech services, and actually to any type of services.

The decision to buy also depends on the *attitude* of the potential user. Attitudes vary from enthusiastic to indifferent to hostile. Everyone has certain opinions and tendencies toward almost all elements of society: politics, art, education, food. These attitudes allow for a coherent response to many diverse subjects. An attitude creates an attractive or rejective environment for a service. For instance, some people may consider that "cyberservices" or, "virtual services," are not serious or reliable enough because they cannot be touched. While others will appreciate them only because they are a breakthrough compared to more traditional solutions.

Past experiences also play a large role in the purchasing decision process. These experiences can contribute to behavioral *learning*. Someone who has been unhappy with an e-shopping service after using it, will have a tendency to turn away from this type of service and will instead consider more traditional services. In addition, a buyer who is satisfied with his IBM computer will most likely prefer an IBM on-line services solution. This preference goes to a brand with which a customer is already familiar.

As R. Pittman, chief executive of AOL, indicated, " I remind my people all the time that Coca-Cola does not win the taste test. Microsoft is not the best operating system. Brands win" [2]. With more than 12 million members, AOL is today the most successful on-line service provider, having bought its main rival Compuserve in 1997 and, so far, dwarfing its other competitors including Microsoft. One of the key success factors of AOL is its uncanny ability to design on-line services that are perceived as easy-to-try and easy-to-use by customers.

2.2.2 Understanding high-tech services business customers

Contrary to consumers, corporate customers are an organization, meaning that the purchasing decision is often made by various people at various levels of the organization.

The purchase of industrial goods and services rarely depends on a single person but usually on a group. In such a group, there are the following participants: the *user,* who needs a good or service and prepares the specifications; the *go-between,* who puts the user in contact with an outside supplier; the *adviser,* who is usually the subject specialist (e.g., in computers, software, telecommunications); the *purchasing agent,* who chooses the suppliers; and, the *decision maker,* who signs the purchasing contract.

The price of a particular high-tech service strongly determines the number of participants in a purchasing group. A subscription to a specific on-line database, worth some thousands of dollars, can be made directly by a development engineer; but an investment in a global network service, or, the outsourcing of the development and maintenance of technology services, representing a total value close to several million dollars will be carefully scrutinized before a member of the executive board will sign the purchasing contract.

The possible application of a service within the firm and its impact on the bottom line are usually the first criteria to be considered in a corporate purchasing decision. However, this decision may also be induced by the environment of the firm. Moreover, the purchasing behavior may also be affected by the organization of the firm. Likewise, the characteristics of each person involved in the purchasing of the service may modify the buying decision (see Figure 2.4).

Let us consider the example of collaborative services to illustrate the interactions of those different elements. Collaborative services include messaging services and groupware services such as document management, group calendaring and scheduling, and corporate directories. The key factors driving the adoption of collaborative services are ease of use and administration, at relatively low costs, which yield high returns on investments because of the new interactions they create both internally, within a firm, and externally, with the various actors of the environment.

2.2.2.1 Environmental factors

Among the many reasons to buy collaborative services solutions are the pressure of environmental factors external to a firm, namely, the economic situation, demand level, competition, technological evolution, and the political context.

In a way, the economic crisis at the beginning of the 1990s has given in to the rise of collaborative services. Faced with a flagging demand, many firms turned to downsizing and rethinking the way they were doing business. One of the elements to implement the major business process re-engineering they had made was to put in place new communication tools between their employees to replace the hierarchical structure that

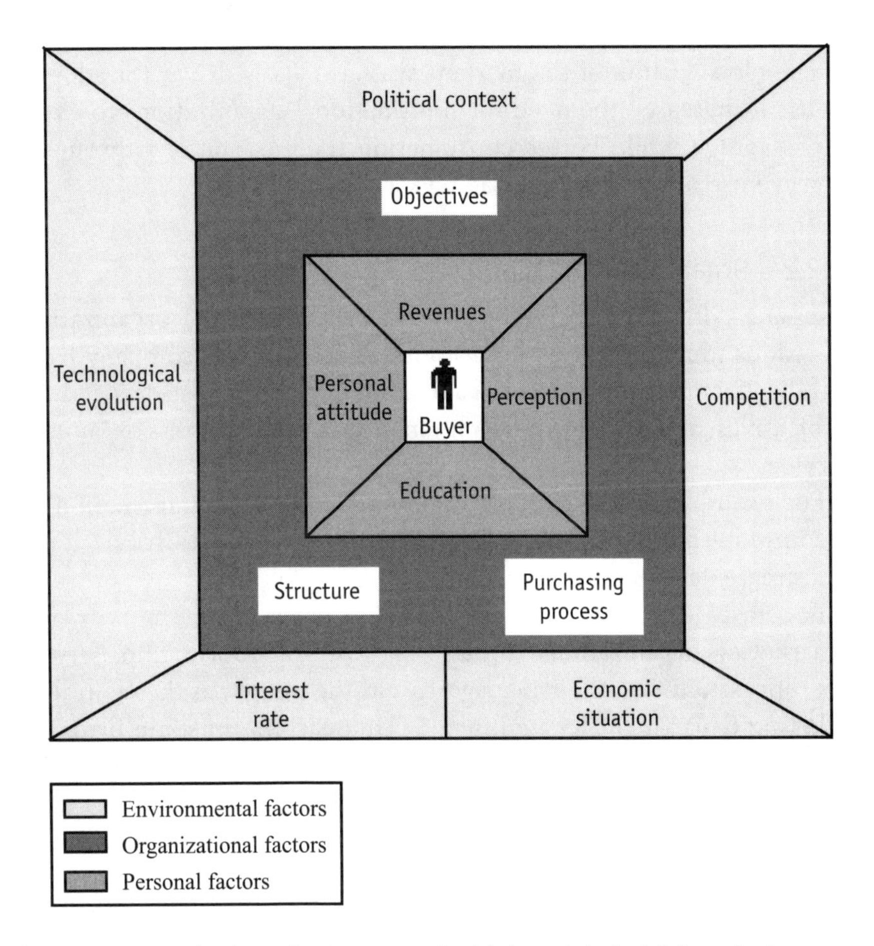

Figure 2.4 Purchasing criteria groups for high-tech industrial products.

has been laid off as well as ease the flow of horizontal information between various departments which have to work together.

In the mid 1990s, with the up-turn of the economical situation, competitive pressures were a key accelerator of the success of the collaborative services. Firms using collaborative services had such a competitive advantage, both in improving their time to market and lowering their operational costs, that their competitors, who were still not equipped, felt compelled to use collaborative services as well.

The globalization of the markets was also a key driver for adoption because it increased the need for multinational organizations to extend their coverage while better coordinating their operations through an extensive use of universal messaging services.

2.2.2.2 Organizational factors

When analyzing the purchasing behavior of a firm, the organizational dimensions must be taken into account as well. Every company has its centralized or decentralized organizational structure, its procedures, its objectives, and its politics which can influence the success or failure of a service.

For example, in the case of collaborative services, information and telecommunication information systems (MIS) managers are no longer the ultimate decision makers when choosing a solution.

For instance, nowadays collaborative services are heavily used in marketing organizations to ease the flow of information between sales representatives, their managers, and the people in the centralized marketing department located in the headquarters. In some firms, the purchasing decision will be made by the MIS department after the validation of the marketing manager. In some decentralized medium-sized firms, the purchasing decision will be made by the marketing manager. Finally, in some major consumer goods firms, the decision will be made by top level management because it is considered to be a key competitive advantage or a core competence to nurture.

2.2.2.3 Personal factors

Since the profile of the decision maker and participant in industrial purchasing groups may vary, it is extremely important to consider the individual characteristics of each one. As seen previously, these are the sociocultural, psychosocial, personal, and psychological factors.

One may think that the purchasing of industrial solutions, being equipment or services, is purely rational, considering only the performance/price ratio. However, different studies show that other criteria, such as the supplier's credibility, service, and long-term commitment to support the product are as important as the sheer performance of the service (assuming the minimum performance level is offered).

However, there is often a tendency to focus on the functional characteristics of the service and its supposed performance, instead of on its ability to fulfill the needs of the customers.

2.2.3 Some specific purchasing criteria for high-tech services: the attitude toward innovation and risk

Because using a high-tech service often means taking the risk of experiencing the initial problems of a novelty, the customer's attitude toward innovation and toward risk must be singled out as specific purchasing criteria.

Many studies have been carried out on the new product adoption process. The results of these studies can be adapted to the world of high-tech services, which is often characterized by a high degree of innovation.

These studies show that not all customers (individuals or organizations) react to new services in identical ways. Some customers will use new services immediately while others will buy them later. The comprehension of the various customers' attitudes is extremely important to jump-start positive feedback leading to increasing returns according to the model introduced at the end Chapter 1 (see Section 1.3.2.3).

Everett M. Rogers, the theoretician of innovation, distinguishes between five classes of customers and characterizes them using psychological traits [3]:

1. *Innovators* have an adventurous spirit; they enjoy trying new products and are venturesome.
2. *Early adopters* are often respected opinion leaders, who are more careful than innovators.
3. Members of the *early majority* like to analyze a product before buying it. This is the case of the nearly 13 million Americans who first began accessing on-line services in 1997. Representing 23% of the users, they reflect a growth in usage by "middle-America"—they are from older age groups, less highly educated, and less frequently from the upper income groups typical of users who have been wired for several years.
4. Members of the *late majority* are skeptics who go along with the majority, but much later. For on-line services, they represent

people planning to get on line. According to a recent survey, these hesitant prospects are a slice of mainstream America: nearly 60% are age 35 or older, 47% have a high school education or less, and 65% have household incomes less than $50,000.

5. *Laggards* are conservatives who do not buy a product until it has become part of tradition.

All five classes are theoretically divided following a normal distribution curve (see Figure 2.5).

In fact, to optimize the introduction of a new high-tech service, innovators and early adopters should first be identified. These two groups of potential customers will give the service its acceptance and will see to the winning over of other customer groups.

Furthermore, studies show that the more complex an innovation is, the more time it will take for the innovative product to be accepted. Indeed the attitude toward innovation is strongly related to the attitude of the customer toward the risk they perceive when deciding to use a service. The risk can be many-sided: it can be about the lack (or absence) of performance of the service; it can be about the financial cost if the service does not work; it can be about the loss of social (or corporate) status associated with a service that goes wrong; or, it can be about the inability of the buyer to use the service correctly.

Actually, what is important to consider is the risk perceived by a given customer. Each person and organization gives a different *importance* to the various kinds of uncertainties that may occur. For instance, the failure of a

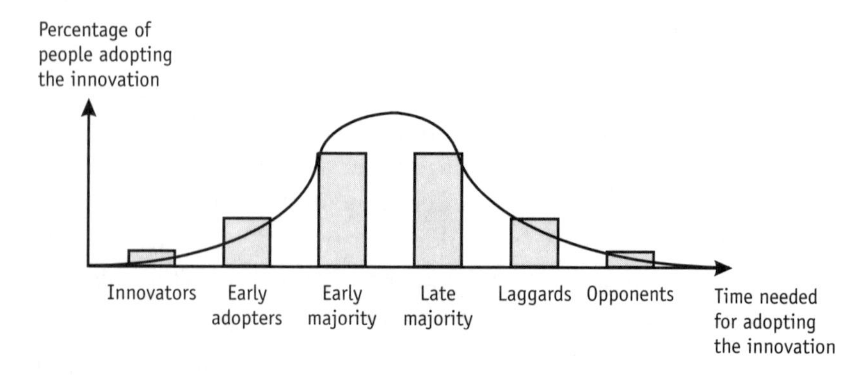

Figure 2.5 The attitudes of customers toward innovation.

network service will be more important for an on-line vendor (which will lose revenues by the hour) than for, say, a university, which can survive for days even without a data network or transaction services. Similarly, each individual has a different perception of the *probability* of occurrence of a given risk. For example, young people always tend to minimize risks and are much more prone to adopt any innovative services while experience shows that senior citizens are much more conservative.

The vast majority of consumers (and businesses) are risk averse and try to reduce risk during the purchase process. This behavior has been identified through many studies [4] but it is most specifically relevant to high-tech services.

On the one hand, services have a higher perceived risk of purchase than goods [5]. The main reason is that a service being an experience cannot be shown first hand to be evaluated by a customer prior to consumption. Furthermore, it is often difficult to replicate one service experience from one customer to another. Consequently, customers cannot easily figure out the quality of the service they want to use, which increases the associated perceived risk.

On the other hand, high-tech solutions, should they be goods or services or a mix of both, are very often disturbing for customers when faced with a technically complex solution of which they do not understand all the elements. But, these customers do realize that this solution is likely to change quickly over time and can suddenly become obsolete. Consequently, high-tech services are usually associated with a high level of perceived risk by potential buyers or first time users.

Looking for security, customers—either individual or corporate—will try to go for what they perceive as the safest solution. They will try to lessen the perceived risk of a given high-tech service through various ways:

- They will be loyal to a brand or a company they can trust and which they know will be around for a sufficient length of time to guarantee the durability of the solution.
- They will search for references, either through word-of-mouth if they are consumers or, in the case of firms, through benchmarking or surveys among their competitors, their suppliers, or other firms which have already used the service.

- They will rely on the advice of opinion leaders, often labeled "gurus" in the corporate world, or technical experts or consultants, as well as any other external source of information such as advertising or articles.
- If possible, they will try to rely on a past experience with a similar service. If not, they will try to have a limited experience through a pilot project in the case of firms, or, in the case of consumers, by taking the opportunity of a free trial offer, if available.

At the prepurchase stage, technology and innovation are not enough to convince the market to adopt a new high-tech service. The service vendor must understand the customers' needs in terms of performance but also in terms of psychological expectation. The vendor must also propose all the occasions for the customers to reduce the risk aversion which inhibits the purchasing behavior of many of them, should they be final consumers or businesses. Companies such as EDS or Atos, for instance, will provide their potential business customers with good references, as well as with free telephone numbers to reach qualified experts and field staff.

The vendor must also try to set realistic customer expectations, through the definition and clear understanding of the requirements that the service is trying to meet and how it is going to meet them. Most specifically for complex services offered to large organizations, there is a need for continuous validation to check that what is being implemented is meeting the requirements that have been defined. For instance, Andersen Consulting systematically sets a "business case" with each customer to define and measure the expectations about a given service, most specifically for integration or outsourcing services. This prevents building "white elephants," huge and complex solutions which bring no value to the customer. This is also the way to build customer loyalty.

2.3 Creating customer loyalty

Once the customer has made the decision to use a high-tech service, the service vendor must ensure that the customer will get a positive experience in line with their expectation. If pleased, the customer will not hesitate to reuse the service, which will increase the vendor's turnover. Like in any service business, customer's loyalty is the only way to a profitable high-tech service.

On the other side, if the customer is not satisfied, it may stop using the service or switch to another vendor, with an immediate and significant impact on the vendor's bottom line. This means that a high-tech services provider must be able to detect and resolve any problems a customer is having, especially if the latter does not communicate this.

2.3.1 The impact of customer loyalty on profitability

Various studies have underlined the importance of customer loyalty in the service business. Reicheld and Sasser [6] have shown that reducing 5% of customer's defections generates 85% more profits in one bank's branch system, 50% more in an insurance brokerage, 75% more in a credit card business, and 35% more in a software vendor.

First, it is obvious that a loyal customer will ensure a regular flow of revenues for a vendor. But, a loyal customer may also contribute to the rise of revenues through referrals. As we have seen, many customers try to lessen the risk when using a high-tech service. For a potential user, one way to do this is by adopting a service which has a good reputation, especially recommended from a trusted source of information. Pleased customers often make positive word-of-mouth comments to their family, their friends, their colleagues, or their peers. By convincing new potential users, they create positive feedback, which will grow as long as the vendor is able to keep its customers satisfied so that they will recruit new ones.

Furthermore, keeping long time customers also helps to reduce the fixed and operating costs. In the case of fixed costs, the more important are often marketing costs to attract new customers, namely, advertising or direct marketing. But high-tech services experience higher fixed costs than average services most notably because of the investments in computer and network hardware and software, as well as retaining a highly educated staff.

Once prospective customers have started using a new service, they belong to the installed base, meaning that the vendor will not have to target them as prospects, and that the marketing effort is less important to produce more sales from these users. In fact, that means a high-tech services vendor should be able to identify new customers, to enter their references in a database, and to regularly monitor that customers are active and have not defected to another vendor.

Furthermore, loyal customers help to cut operating costs because they are used to the procedures and services. They know the role [7] they have to play so that the service will perform effectively. The role can be limited to providing some specific information on a given date. But it can also be personal interaction with some sophisticated operations through various screens and menus, like in the case of some electronic commerce applications. In all cases, experienced customers have fewer problems than first-time users, meaning they have fewer questions for the support call center.

2.3.2 How to gain and retain a customer?

There are different ways to keep customers devoted to a service. The first one is to interact with them during the service, which can be made in a creative and original way through the electronic channels that support many high-tech services. The second way is to aggressively manage the defection process. Finally, a sound guarantee policy is also an efficient way to retain customers.

2.3.2.1 Interacting with the users

On the Web, nurturing customer loyalty can be done on an ongoing basis by *interacting with electronic users*. This technique of direct interaction with consumers is at the core of "relationship marketing," which acknowledges that the goal of marketing is no longer to achieve one single transaction with one customer and move on to the next one, but to build a strong and long-term relationship with existing customers [8].

To achieve such a relationship, service vendors must know their customers and their needs almost on a one-to-one basis, in order to deliver the best adapted solution, with the requested quality. They also have to identify and quickly answer any customer's complaint in order to maximize their retention of customers. They must intensify their contact with their customers in order to serve them well.

Accordingly, for the service vendors who implement "relationship marketing," the capacity to obtain and apply customer information within processes is a key strategic issue. This often places the company in the position of requiring sensitive personal information from customers. But the advances in information technology have fundamentally altered the channels through which companies and customers maintain their relationships. Because they can be used repeatedly and interactively, information technologies create a new type of bond with customers (or between employees in the case of the intranet). Given the way in which people share their favorite Web sites (the places they like to visit and the electronic communities to which they belong), electronic customers are no longer passive targets, but active members of a virtual network or community.

In contrast to traditional marketing programs that are essentially one-way communications to relatively large audiences with varied buying interests, electronic marketing on the World Wide Web may develop an affinity more easily between the customer and seller through a closer, more intimate relationship. Here, customers have the ability to get information on demand that relates precisely to their exact needs and develops a strong feeling of being highly served by the selling company. Since customers and prospects alike want to feel a personal relationship with the companies providing services for their needs, the concept of being able to get information in a personalized manner and to communicate interactively through e-mail or other electronic means, dramatically enhances the company's image in the buyer's mind.

The desire to establish long-term customer relationships with increasingly sophisticated demands has led companies to seek new ways of acquiring, managing, and utilizing customer information.

Companies such as A. C. Nielsen and Internet Profiles Corporation (I/PRO) are currently developing a new system to measure how many people are coming to specific sites, who they are, and how much time they spend at that site. The system can be used for analysis of the server logs for measuring traffic patterns.

A more aggressive technology employed to track customer information is a technology called "cookies." This mechanism is fairly simple and extremely helpful in enabling a Web server to understand which user with which browser is speaking to it. Corporate Web sites use cookies to make it easier to execute legitimate business processes, like keeping on-line orders straight. The term "cookie" actually has been in use in computing for decades to describe a token or ticket that is passed back and forth between programs or routines within a program. The New Hacker's Dictionary likens it to the mundane coupon you get from the dry cleaners: when you return for your clothes, the dry cleaner matches up the tickets to ensure you get back the correct items.

The need for Web browsers and servers to exchange cookies is due to the stateless nature of hypertext transfer protocol (HTTP). Unless something special is done, Web servers are only aware of users when a transaction—sending or receiving information—is in process. The moment the transaction is complete, the server forgets about the user and makes no attempt to correlate subsequent transactions with previous exchanges.

Cookies have obvious appeal for corporate webmasters because they deliver a fairly standard infrastructure to compensate for the stateless nature of HTTP. This is essential for anything beyond the most rudimentary on-line order system, to make it possible for users to customize their interaction with large complex Web sites, and to customize the way they view a Web site.

Netscape Communications and Amazon Books use a temporary cookie to maintain state information for shoppers. This enables a site to spread products and information over multiple pages, or to put the order entry forms on a separate page. As the user selects products they want to buy, the server indexes these selections to the session key carried as a cookie by the user's browser.

Persistent cookies—those stored on the user's file system—also provide a convenient location to store user preferences that are likely to be used each time the user visits a Web site. Netscape used a simple cookie to enable a user to set a preference for viewing its site with or without frames. Another cookie and Javascript code is used for Netscape's personal page service. Search sites such as Excite and Search.Com also use cookies to customize what kind of information the user prefers to see when they log onto the site.

2.3.2.2 Managing customer defection

Managing customer defection is of importance for a high-tech services firm where the defection rate can be quite high—about 35% per year in the cellular phone industry and more than 50% annually in the U.S. pager industry.

Most of the defection management process in high-tech services deals with the identification of the defector. In more traditional people-based services, a potential defector can be identified on the spot by an employee if something goes wrong. Similarly, a human interface is often used to reduce defection in high-tech services, particularly for business-to-business services. However, this is not always the case in information-based services where a customer can have problems logging onto a server, has to wait too long, does not understand all the operations requested to be performed on a given screen, or finds that the proposed service is too expensive.

In all those cases, and there may be others, the user will quit and may decide not to complain, but will never again use such a service. Consequently, because of the absence of the human interface, the recovery process must be designed in advance and embodied in the electronic service through a variety of help functions and automatic procedures which appear automatically on the screen when something goes wrong to help the user sort it out. Ultimately, a hotline number should always back up the program and should be clearly presented to the customer.

2.3.2.3 Offering service guarantees

Another efficient way to keep customers is the service guarantee and, more specifically, the unconditional service guarantee. It is an explicit guarantee which is not attached to a specific performance. It promises customer satisfaction with a refund or free problem resolution should any difficulty happen.

For customers, such a guarantee tends to decrease the perceived risk of a service while in most cases, the image of reliability of the vendor is increased. It also makes the customer more comfortable in using and reusing the service.

For a high-tech services firm, an unconditional guarantee can provide a significant way of differentiation. However, the preliminary condition is that the firm must be perfectly ready as far as its operations are concerned. If not, it may end up with serious financial difficulties (even bankruptcy in the event it has limited cash) if all the customers ask for a refund.

In the case of corporate customers, which are using various sophisticated services, the key to obtaining customer loyalty is both in the quality of the solutions provided and the existence of an organization dedicated to the customer. Such a team includes not only the technical support staff, but also professionals who are in charge of constantly monitoring the relationship with the customer.

Corporate customers are managed by an "account manager" or, a "client partner," who visits customers on a regular basis, assesses their satisfaction and plans the necessary actions should something need to be corrected. "Such a team is also the best way to constantly improve the vendor's knowledge of the characteristics of the customer's business, another key element of keeping customers loyal," noted a manager from Atos.

To ensure that their customers are satisfied, some vendors such as Andersen Consulting are even ready to make a direct connection between their fees and the value they are bringing to customers. Such is the case of the service Andersen Consulting provides to Mercedes Benz in managing the plant information system that manufactures the new Smart microcar where Andersen Consulting's fees depend on the number of cars manufactured.

In other cases, it is not uncommon to see a joint venture between the vendor and its customer to strengthen the relationship. For instance, Capsam Consulting is born from a joint venture between Cap Gemini and the Finnish insurer, Sampo, while Connect 2020 is a venture between Andersen Consulting and the British utility company, Thames Water. For the service provider, such teaming effectively locks in the customer as a counterpart with the financial risk.

2.3.3 Evaluating customer satisfaction

High-tech services firms need to evaluate the degree of efficiency of their customer loyalty programs. To do so, they rely on different ways to measure customer satisfaction.

In business-to-business, the major vendors:

- Track the payments (since the payments are usually attached to the official acceptance of a service or of a significant part of a service by the customer. If the customer does not pay, it can be attributed to dissatisfaction with the solution delivered);

- Trace the evolution of the turnover with a given customer and the number of new projects "in the pipeline" waiting to be negotiated. A negative trend is a clear sign that the customer is not pleased and has found other ways to fulfill its needs;
- Monitor the performance of answering complaints, both in quantity (e.g., average lead time to answer a given type of complaint) and in quality (was the problem definitely fixed or not?);
- Use customer satisfaction surveys. Some vendors survey customers on a regular basis, with a minimum of once a year. But IBM Global Services also has a survey performed by an external source at the end of major projects.

Indeed, a vendor can always check all the technical aspects which may be part of customer satisfaction such as service availability, response times, defect rates, etc. and all the elements which are part of a quality service. As we will see in Chapter 3, these criteria need to be carefully monitored and constantly improved. But the performance of each of these components does not guarantee customer satisfaction [9].

In some cases, performance indicators may be correct while a potential problem may exist that will strike the vendor in the future. The reason is that what is important is the customer's perception of the service and the comparison of the performance with its expectations.

Accordingly, it is necessary to talk to the customer to gauge customer satisfaction. Giving users the opportunity to voice dissatisfaction helps the vendor to anticipate potential problems and to put in place adequate corrective actions; it may also prevent users from looking for another vendor or solution once they feel their problem has been correctly understood and will be fixed rapidly.

Generally, the customer is surveyed to evaluate the service through a written or a telephone questionnaire, or in a face-to-face interview. The method depends on the number and the size of the customers to be examined. Whatever the method, some basic rules have to be observed so that the survey will be efficient (see Figure 2.6):

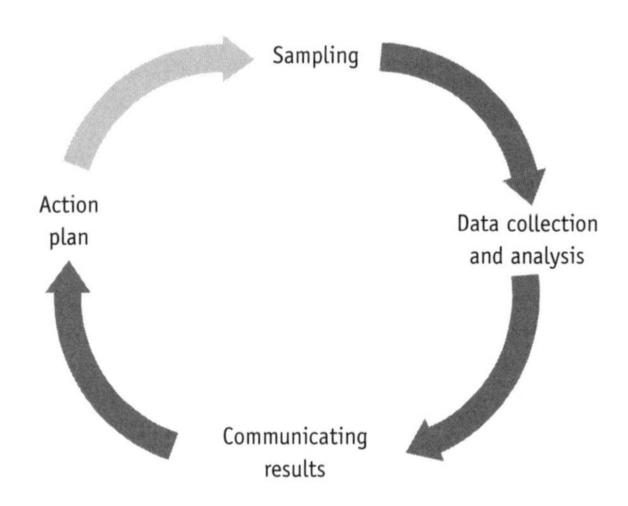

Figure 2.6 The four key steps in a satisfaction survey.

- The sample of users surveyed must be adequate and should not have a bias. Such a case occurs when one specific group eclipses others or when the same set of customers is polled continuously. In the latter case, it has been observed that if a user is interviewed and then sees some improvement in the service, this perception influenced the answer in the following survey. Consequently, any improvement in the service should be checked with untried customers (unless the improvement is specifically targeted to a precise category of users).
- The format of the data must be simple to collect and analyze so that a quick action plan can be designed and implemented. In some cases when there are few customers, the information to be measured can be decided up front. Such is the practice at Cap Gemini, where, according to one of its managers, "the indicators of satisfaction are first defined with the customer. For example, it may be the technical quality of an application software or the level of competence to be transferred to the users. Then the customer satisfaction is measured through surveys and tracing all the events and actions on a project."
- The results must be communicated openly to all the participants. Most specifically, it is useless to hide the results if they are bad;

users already know the results, since they provided the information about poor performance. For instance, so far, user satisfaction with on-line services remains low. Users' overall satisfaction with their experience is most highly correlated with their degree of satisfaction with their Internet or on-line service provider. What they would most like to see improved is speed of access, followed by reliability of connection.

- An action plan has to be devised to solve any problems pointed out in the survey. The actions should be designed in association with the users who asked for them; this ensures the users' engagement while giving credibility to the vendor because the vendor has followed up with the user. This is not always the case. Sometimes, customer surveys are just a collection of data with no results. If the vendor conveys this to the users, they may feel frustrated at not being listened to, and the survey may create or reinforce a negative perception of the vendor and the vendor's services!

Whatever the phase in the customer relationship life cycle, we have seen that customer satisfaction is at the center of any sound competitive strategy for a high-tech services firm. This is an absolute necessity and must be engraved in the minds of all employees, should they be in contact with customers or designing a new service.

At IBM Global Services for instance, customer satisfaction is one of the personal goals of each collaborator, employee, or manager. But this is not enough. Leaders must constantly show the importance of being in touch with customers. They visit business customers with their account executives to stay in close touch with the market. They make sure that front-line people, interacting face-to-face or by telephone, have the right information about the customers.

Ultimately, successful service vendors integrate all the knowledge about customers in the definition of their service standard [10]. They start identifying customers' needs as well as expectations, and decide whether or not they fit in the corporate strategy.

If the answer is positive, they assess the market environment and benchmark practices worldwide, both in and out of the industry, in order to anticipate customers' expectations. For example, some high-tech services users who travel extensively have seen the latest innovations and their expectations in this area are heightened.

Once the services standards are outlined, they are confirmed again with customers through focus groups. By doing so, service vendors make sure that their service specifications will represent the key features that drive the satisfaction of their customers.

2.4 Chapter summary

In any business, the key to strategic success is to bring value to customers. Value is what the customer gets for his or her money, that is, the benefits provided minus the cost of acquisition. This value will be different for each customer according to his or her expectations and personal experience, the comparison of service offers of competitors, the vendor's reputation, and ultimately the service performance.

In the service business this value is not brought by a product, since the customer is mostly buying a performance, an intangible result. Consequently, the key to success in the high-tech services business does not lie in the aggressive management of the product life cycle as in the case of high-tech products, but in the effective management of the customer relationship.

One can identify three different steps where the customer's behavior will vary significantly toward a high-tech service: prepurchase, consumption, and postpurchase.

To achieve success in the management of the customer relationship life cycle, a service vendor must first have an extensive understanding of the customers, not only their identification, but also their needs and decision-making processes to make them select the vendor's service offer. For that matter, the vendor has to differentiate individual consumers from businesses and organizations. The example of using on-line services illustrates how consumers are influenced by four classes of factors: sociocultural, psychosocial, personal, and psychological.

First, corporate purchasing decision makers consider the possible application of a service within the firm and its impact on the bottom line. But, this decision may also be affected by the environment of the firm, the organization of the firm, and the characteristics of each person involved in the purchasing of the service. The case of collaborative services illustrates the interactions of these different elements.

Finally, the customer's attitude toward innovation and risk must be singled out as specific purchasing criteria because using a high-tech service often means taking the risk of experiencing the initial problems of a novelty.

Once the customer has made the decision to use a high-tech service, the service vendor must create customer loyalty, which is the only way to achieve a profitable high-tech service. This chapter considers the different ways to keep a customer devoted to a service. The first way is to interact with the customer during the service, which can be done in a creative and original way through the electronic channels that support many high-tech services. The second way is to aggressively manage the defection process. Finally, a sound guarantee policy is also an efficient way to retain customers.

In addition, high-tech services firms need to evaluate the degree of efficiency of their customer loyalty programs in measuring their customers' satisfaction. To do so, they track payments, trace the evolution of turnover with a given customer and the number of new projects "in the pipeline" waiting to be negotiated, and monitor the performance of answering complaints. They also use customer satisfaction surveys.

Ultimately, successful service vendors integrate all the knowledge about customers in the definition of their service standards so that the service specifications will represent the key features that drive the satisfaction of the customers.

References

[1] Viardot, E., *Successful Marketing Strategy for High-Tech Firms*, 2nd ed., Norwood, MA: Artech House, 1998.

[2] Gunther, M., "The Internet is Mr. Case's Neighborhood," *Fortune*, March 30, 1998.

[3] Rogers, E. M., *Diffusion of Innovations*, New York: The Free Press, 1983.

[4] Bauer, R. A., "Consumer Behavior as Risk Taking," *in* R. S. Hancock, *Dynamic Marketing for a Changing World*, Chicago, IL: American Marketing Association, 1960, pp. 389–398.

[5] Guseman, D., "Risk Perception and Risk Reduction in Consumer Services," *in* J. H. Donelle and W. R. George, *Marketing of Services,* Chicago, IL: American Marketing Association, 1981, pp. 200–204.

[6] Reicheld, F. F., and W. E. Sasser, Jr., "Zero Defections: Quality Comes to Services," *Harvard Business Review,* September–October 1990, pp. 105–111.

[7] Soloman, M. R., et al., "A Role Theory Perspective on Dyadic Interactions: The Service Encounter," *Journal of Marketing*, Vol. 1, No. 49, Winter 1985, pp. 99–111.

[8] Christopher, M., A. Payne, and D. Ballantyne, *Relationship Marketing,* Oxford: Butterworth-Heinemann, 1991.

[9] Hallows, R., *Service Management in Computing and Telecommunications*, Norwood, MA: Artech House, 1995.

[10] Tocquer, G. A., and C. Cudennec, *Service Asia. How The Tigers Got Their Stripes*, Singapore: Prentice Hall, 1998.

3

Quality for High-Tech Services

Besides customer focus, quality is the second lever used to achieve success in any high-tech services business. In an era where customers are hungry for quality, offering a high-performance, fault-free service is a powerful way to differentiate from competitors. This helps to keep existing customers by effectively delivering the performance which has been promised. Another advantage is reduced anxiety and confusion because customers get what they expected. Likewise, it helps in attracting new customers because the image of a quality provider—backed by reality—lowers the perceived risk for a potential user.

Geoff Unwin, one of Cap Gemini's vice-presidents, said, "Quality is essential for our customers and for ourselves. Indeed, through quality, we are going to revolutionize the world of services (the way the Japanese revolutionized it) for the car industry."

But what exactly is quality? Another manager at Cap Sogeti gives a very clear and straightforward definition of quality: "Quality is a contractual commitment that sets up the various levels of service and performance provided to the customer."

Such a definition is very similar to the one given in product manufacturing where today quality is often characterized as "delivering the right product to the right customer." However, there is a significant difference. Actions to better product quality have historically concentrated mostly on the removal of product failures. First, manufacturers implemented strict inspections of all the products at the end of the manufacturing line. Then, they moved one step ahead by improving the quality in the manufacturing process to get it right the first time. Finally, they considered the whole supply chain and ameliorated logistics, such as the storage, distribution, or billing facilities to ensure a smooth delivery of the expected product to the customer.

High-tech services vendors have experienced a similar evolution. Indeed, a significant part of the service quality relies on the elimination of defects before the customer sees the service, by focusing on the design and the implementation of the service (see Section 3.1).

However, this is not enough, because as we have seen previously, a high-tech service is an experience, "something" which happens at a given time. To ensure the repeated quality of such a performance, service quality cannot be a distinct agenda to be completed. It has to be an ongoing component of service operations.

Consequently, high-tech services vendors are developing and using a variety of specific measurement tools to constantly monitor their performance versus their customers' expectations and perceptions. As we will see in Sections 3.2 and 3.3 of this chapter, service quality is achieved differently for business customers (see Section 3.2) and consumers (see Section 3.3).

3.1 Quality process design and implementation

Failures in service quality are not limited to technical problems impeding the expected performance of a given service. A poor image of quality for a high-tech service comes more generally from the gap between customers' expectations of a service and their perception of the service provided [1]. Quality oriented high-tech services firms intend to close or decrease this gap. To do so, they must also lessen other differences (see Figure 3.1), namely:

- The difference between the users' expectations and the specifications of the service;
- The variance between the specifications of the service and the actual service delivery;
- The gap between the service delivery and its perception by the customer;
- The gap between customers' expectations of a service and their perception of the service provided.

3.1.1 The difference between the user's expectations and the specifications of the service

One of the most significant causes of problems in the quality of high-tech services is the discrepancy between what customers require and what managers and designers believe customers require. Many service providers think they know what their customers look for, but experience shows that they are very often off the mark.

As seen in the previous chapter, customers' behavior is not very easy to discern especially when facing the rich complexity and unfamiliarity of a high-tech service. Such an experience may involve subtle interaction

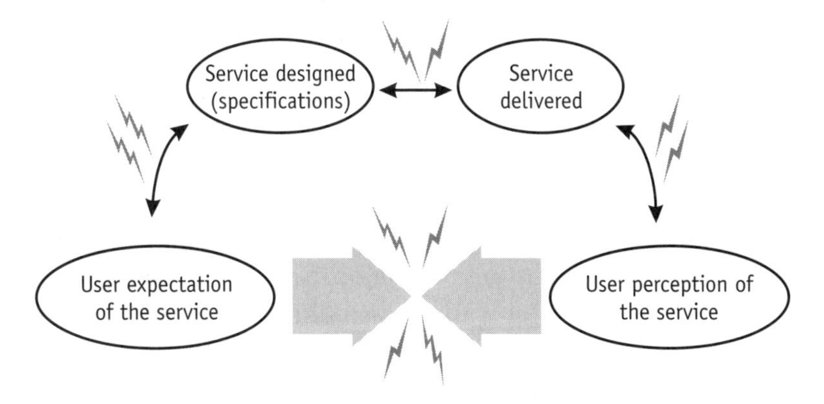

Figure 3.1 The various dimensions of quality problems for high-tech services.

between the users and the service, (which cannot always be easily summarized). However, our experience is that service designers and their managers are not always keen enough to listen to their customers.

The only way to ensure that quality of service will be achieved at this stage is to include the customers' desires in the specifications of the service. This requires first-hand, extensive knowledge of customers' expectations. In business-to-business services, such knowledge is captured and validated during the pre-sales phase when qualifying a project and reviewing the bid. In business-to-consumer services, market surveys, focus groups, and test customers help the service providers to gain a better understanding of the actual customers' expectations that will have to be fulfilled in the service to be designed. Customer satisfaction surveys help also to identify gaps and to make the necessary corrections.

Moreover, one of the beauties of information-based services is their user interactivity. A very clever and efficient way to level the discrepancy between expectations and specifications is to empower the customers by giving them the means to answer their own needs in real time [2].

Such is the strategy of Intuit, which has introduced Quicken Live, a new on-line service that complements the family fund management software, Quicken. Quicken Live allows the more than 10 million Quicken users to specify the program to their own needs, dramatically improving customer satisfaction. Last but not least, through a feedback system, Intuit monitors the various customer requests and selects new ideas for future applications.

3.1.2 The variance between the specifications of the service and the actual service delivery

Although the customers' needs may be correctly anticipated in the service specification, a quality problem may still exist because of the difference between the designed service and its perceived delivery.

In the case of sophisticated business services, such a quality gap generally comes from the employees both in terms of willingness and ability. Overworked, underpaid, or frustrated employees will be less motivated and efficient in providing the needed performance at a given time. (Because most of the operations are automatic in information-based services, the negative influence of employees on performance is mostly felt in the delay of operations, such as the writing of a new software program, the implementation of a new application, or the technical support to customers.)

Other employees may be unable to perform the service according to specifications because they are not qualified enough, or lack training. They may also lack the technological resources and the support required to fulfill their role. Finally, the organization may impede employee efficiency through excessive control, limiting the ability of personnel to quickly answer customer demands.

Consequently, filling this quality gap is dependent on the management of people and the service organization, the third success factor in the high-tech services business, which we will discuss in Chapter 4.

3.1.3 The gap between the service delivery and its perception by the customer

The gap between service delivery and customer perception occurs frequently with Internet-based services, where there are no human interfaces, no contact employees, and no front-line personnel. The difference between the delivery of a service and its perception lies in the behavior of the user when facing the computer screen. In some cases, a user does not have a positive experience because the user does not behave adequately. In other cases, a customer may have a negative perception because of confusing questions put into the service specifications by the service provider.

Indeed, Steve Casey, the president of AOL, made a point when he said, "There is a syndrome inside Silicon Valley that is out of touch with what consumers want [3]." Very often, Web services are not user friendly enough: screens and instructions are difficult to read, and menus are complex and take time to appear. Even when the service is carefully designed, the user can act incorrectly by pushing the wrong button, typing the wrong word, or clicking the wrong icon. Any of these incidents can contribute to building a negative perception of the service making the nonspecialist user at best, frustrated, and at the worst, so disappointed that he or she leaves the Web site with no intention of ever returning.

The solution to achieving superior quality in this area is by designing easy-to-use applications with specific training menus, enticing the consumer to get more experience at no risk and as much support as possible (on screen and off screen).

Another source of a potentially negative perception lies in some Web sites that request confidential information that has no connection to the service being used from the consumer. If requesting such information makes sense from the vendor's perspective (for example, to get a better understanding of users), one must realize that such a demand may trigger some unwillingness in customers who resent being scrutinized.

3.1.4 The gap between customers' expectations of a service and their perception of the service provided

This "credibility gap" or "promises gap" is not very different from that experienced by goods manufacturers because it is driven by the communication of the firm. Overselling to a customer—in a sales pitch or in an advertising campaign—will inevitably lead to the breaking of promises and the loss of confidence in the service vendor.

One must note that it is easier to oversell services than products because, in most cases, the former are intangible, making any pledge easier. As one partner of Andersen Consulting noted: "In our business, selling is easy; what is tough is delivering." The solution to such quality problems is to control communication, and make sure that the communication department is in touch with the reality of the operations, as we will see in Chapter 6.

In the business-to-business market for large high-tech services, Andersen Consulting has found a simple and effective solution: the team that sells the solution to a business customer is also the one that delivers it. This means that the team is committed de facto to any engagement made to its customer and will deliver a quality solution according to the definition given in the introduction of this chapter.

Ultimately, when considering the quality of a service, the customer evaluates not only the performance (as for a finished good) but also the process. For instance, a customer visiting a Web site for electronic shopping will not only judge the number and quality of the products shown on the site but will also notice if the program is easy to use, if it offers attractive pictures and clear explanations, and if it does not request too much personal information to place an order.

As a consequence to achieving perfect quality for the services they are offering, high-tech services vendors must not only rely on the various development methodologies and techniques to build error-proof application software; they must also design operating procedures to ensure that the quality offered is the one expected by the customer. The tools and procedures will vary significantly according to the nature of the customer, whether a corporate customer or a consumer.

3.2 Achieving quality of high-tech services for business customers

To ensure that they are offering top quality solutions to their corporate customers, high-tech services providers are investing a lot in the development of tools and methodologies. But this is not enough. They must also make sure that they are making the right offer to each customer so that the service will match customer expectations as closely as possible.

3.2.1 Developing and using fool-proof software and solutions

The quality of high-tech services relies extensively on the use of state-of-the-art technology, which allows for maximum performance but is not without risk. Consequently, achieving the quality for those services requires not only a careful software development process but also active risk management and the willingness to constantly improve performance.

3.2.1.1 The software development process

Developing software is a risky business. According to a recent survey conducted in the United States [4], about 30% of in-house software projects never materialize, 50% cost more than double the estimated budget, and only 9% make it on time and on budget (see Figure 3.2).

Furthermore, the risk of failure is strongly correlated with the size of the software project. The bigger it is, the more risky it is. More than half of big software projects fail. On the one hand, this is good news for

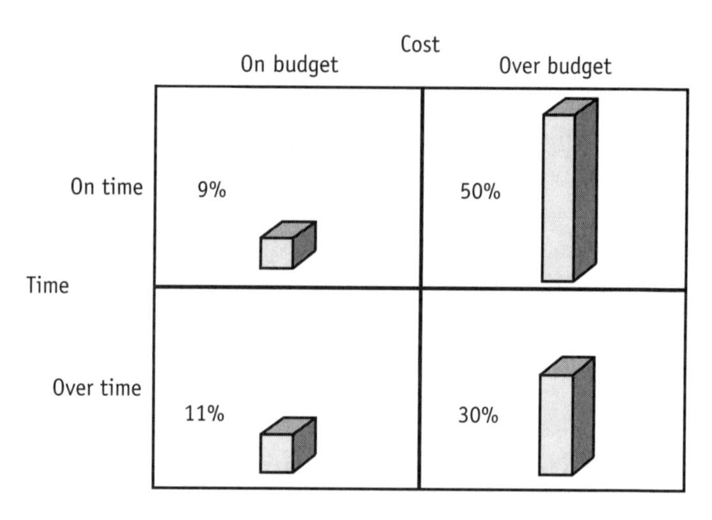

Figure 3.2 Success rates in achieving software projects on time and budget (percentage of number of projects). (*Source:* Software Engineering Institute.)

high-tech services vendors because such a failure rate has pushed a lot of corporate customers to outsource their project software. On the other hand, such figures illustrate the difficulty of achieving the development of software solutions that constitute the high-tech services infrastructure.

Most of the problems come from a lack of control over the software development process through poor analysis of actual needs, a deficit of user involvement, faulty management of risk, and blind faith in technology (for example, computers, operating systems, or programming languages).

In order to counter these difficulties, all high-tech services vendors are using various development methodologies in addition to state-of-the-art technology because they know from experience that too often the leading edge of technology becomes the bleeding edge!

The main evolution has been to move the *model* of software development from a linear model to a circular one with many feedback loops.

For many years, the development of software was sequenced linearly as a series of various steps to be performed from the analysis of users' needs to the technical achievement (see Figure 3.3). The analysis of needs identified the inputs, the outputs, and the transformations to be made. The conception (or design) phase was dedicated to accomplishing a

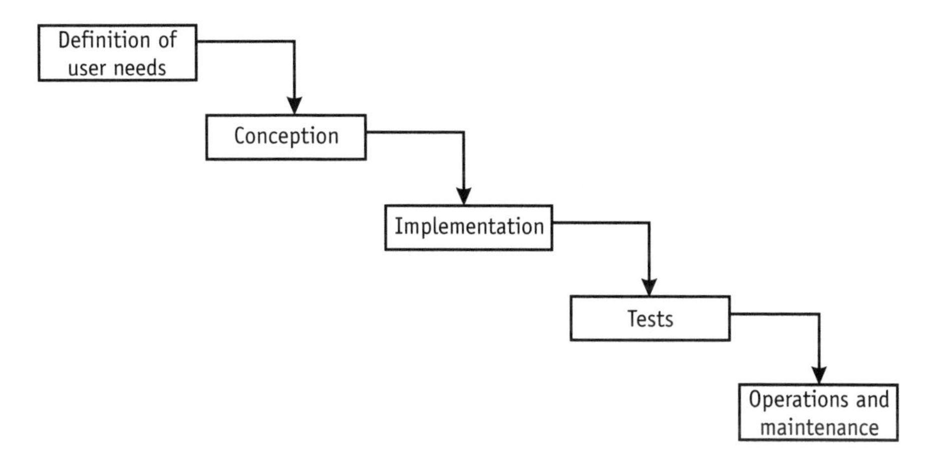

Figure 3.3 The sequential linear model of software development.

detailed technical specification describing files, algorithms, and output documents. The implementation phase was dedicated to coding the program, and the test phase was dedicated to tuning and validating it.

The main downside of this model was its incapacity to take into consideration complex projects and systems involving numerous applications interacting with each other. Moreover, because the model was linear, it did not control the whole process very well. For instance, the testing phase was related only to the global application and never (or rarely) to the fulfillment of user needs.

Consequently, high-tech services providers implemented a new model known as the "V model" (see Figure 3.4). It introduces the concepts of systems and subsystems (components), which gives more coherence to the conception phase by bringing in hierarchy and building blocks (modules) and allows improved management for developing complex application software.

Furthermore, the model adds better control of the whole development process through an explicitly specified set of tests. A validation step of the whole system is introduced and clearly differentiates between the internal validation (or logical coherence) of the software and its fit with the user's needs (validation).

One should remark that this model has some structural flaws. One of the more important is bottom-up checking, starting with the check

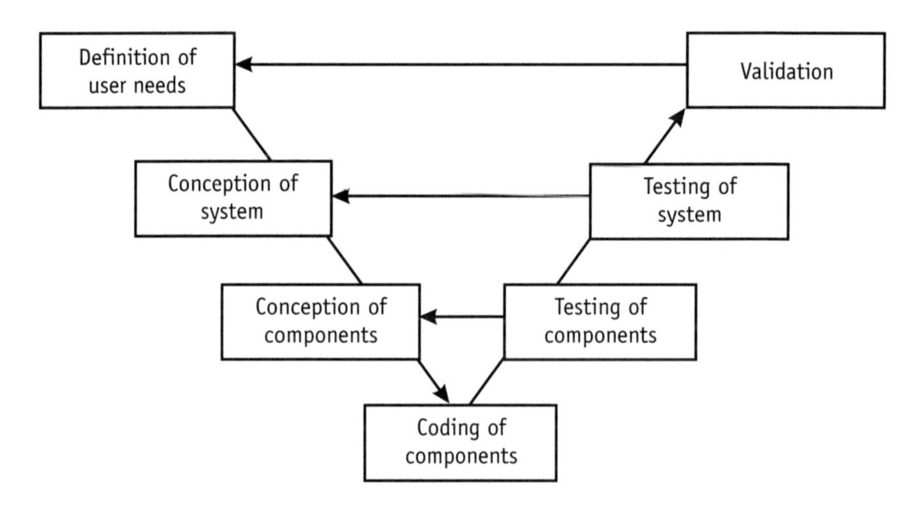

Figure 3.4 The circular or V model of software development.

of components to the check of the system, and finishing with validation according to user needs. Sometimes, it may be very costly to figure out the inadequacy of the software after it has been developed. Though some other models have been proposed, they have more theoretical than practical value. As a consequence, the V model is still the preferred way used by successful service providers to develop software programs on business-to-business projects.

To ensure the technical quality of their software development, high-tech services vendors are also building software using a strict architecture to ensure easy evolution of software programs within services that may have a 10 or 15 year lifetime while taking advantage of the latest improvements in software programming like object-oriented analysis (OOA) or object-oriented design (OOD).

High-tech vendors are also using program generators and dedicated application software to automate a lot of repetitive and low value operations such as software code writing, testing, and documentation. As an executive from Cap Sogeti observed: "Today's ruling technology—the object-oriented approach—facilitates the re-use of components. Those components are modules, which bring together functions and data. To create a new application, one selects an existing structure and modifies its behavior by assembling components. Today, these building blocks are known as ActiveXs, Applets, or Plugins."

But high-tech services vendors have also engaged in the dynamic management of risks.

3.2.1.2 Software risk management

To improve the quality of their software, most high-tech services vendors aggressively manage the risks in developing software projects. They try to identify the risks, to anticipate their potential impact, to make contingency plans, to follow up throughout the project, and to wrap it up in order to build experience.

Such a process costs time and money but has been shown to be the sole way to minimize the risk of errors in software design, which may have bigger consequences, both financially and in overtime, if the software fails when customers use or need it.

Most of the risks can be identified by using various techniques to evaluate the development process. The most famous and widely used technique in the United States is the Capability Maturity Model (CMM) developed by W. Humphrey of the Software Engineering Institute (SEI) for the American Department of Defense in 1989. Similar to CMM is Bootstrap, which is commonly used in Europe, and Trillium, which is preferred in Canada.

All these methods are based on the same principles, namely, a mechanism to assess the capacity of each software development process to achieve the required performance in quality, efficiency, cost, and delay by checking various technical and managerial criteria through a questionnaire. This diagnostic is similar to a process benchmarking adapted specifically to the development of software. It identifies both the soft spots as well as the best demonstrated practices on a project.

The evaluation measures the "maturity" of each process on a scale allowing the organization to map its ability to implement the software development process (see Figure 3.5). For instance, CMM scales any project from 1 to 5:

1. *Initial:* The software project is "ad hoc"; few processes are defined.

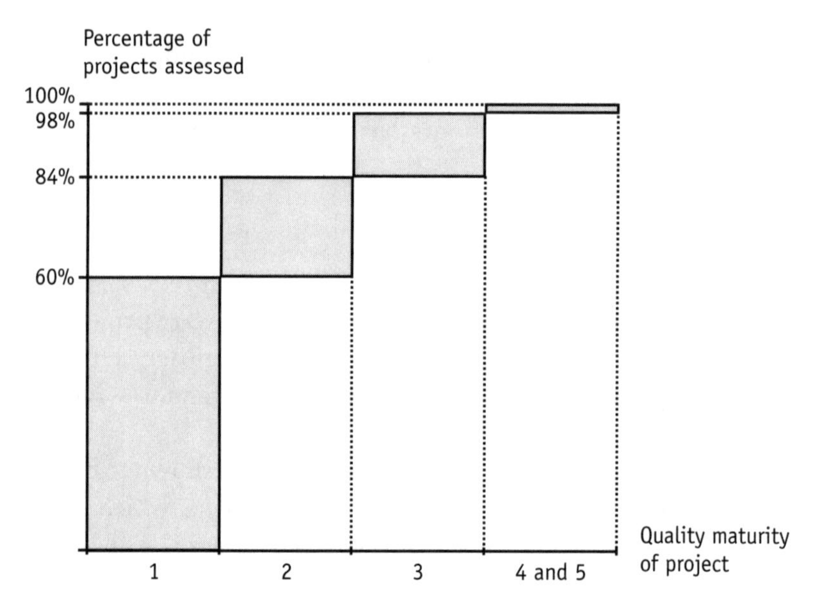

Figure 3.5 Projects split by level of quality according to the CMM method on a sample of 3,000 projects from 600 organizations between 1992 and 1997. (*Source:* Software Engineering Institute CMM.)

2. *Repeatable:* Basic project management processes are established with schedule and functionality; the process can be reproduced.
3. *Defined process:* There is a standard consistent process; it can be generalized to the organization.
4. *Managed process:* The process is controlled and predictable.
5. *Optimizing process:* There is a continuously improving process updated on a regular basis to include technological innovations.

Because each method has its own evaluation process, though the principles are similar, making comparisons is difficult especially on a worldwide basis. For instance, it is not easy to compare the outputs from CMM and Bootstrap because they do not have the same format.

To avoid such problems, in 1998 the International Standards Organization issued a new specific standard for Software Process

Improvement and Capability Development (SPICE), which allows a comparison of results from the various assessments methods customarily used today. More than 25 countries have been involved in the design of this standard and more than 200 project evaluations have been performed so far with SPICE.

3.2.1.3 Improving the whole solution performance

Improving the software development process is the only way to ensure the quality of software projects whose size and complexity are getting bigger everyday, while making them within budget, on deadline and, consequently, improving the customer's satisfaction. It is also a critical factor to achieve productivity and to keep costs under control.

At Bull Computer, the improvement of software process design started in 1992 and led to a 20% decrease in the number of failed projects on customers' sites, as well as to cost savings on maintenance worth six times the value of the investment for the improvement of software development. Another appraisal of the return on a seven person per year investment for inspecting software corrections gave a savings of 47 people per year (not including the increase of customer satisfaction), meaning a net gain of 40 people per year.

The key success factor in improving software development is first the existence of a pool of experts who can make objective appraisals, formulate accurate recommendations, and help to implement them. These experts not only work in house, but also can be put at the customers' disposal.

A second element of success is the establishment of transmitters to ensure that these methods are used in the day-to-day activities of software development units and are fully assimilated by the developers.

Last but not least, perseverance and endurance are a must because software improvement is a continuous ongoing process. Experience teaches that whenever a software development team eases off before it has reached a sufficient maturity level, its quality performance sharply declines.

In any case, involving the end-user in the development to ensure that the software fits user needs is also a crucial element of success.

3.2.2 The right offer to the customer: the procedure for quality processes

The quality of high-tech services is not guaranteed by the quality of various software and processes that comprise the solution offered on the market. Because service is a performance and not a product, it is important to be sure that the solution that is delivered is exactly in line with the needs of the customers. For them, it is the only criteria they will consider for evaluating the quality of a service. If a high-tech service is technically flawless but falls short of customer expectations, it is a poor quality solution.

The 1980s saw many "white elephant" projects involving a sophisticated amount of technology in hardware, software, and telecommunications that ultimately failed to bring any value to end users. Actually, one of the driving reasons why corporate customers started to outsource the making and running of many services to external vendors was the inability of their own information systems departments to deliver quality solutions. Examples can be found in many businesses from banks to insurance companies, from manufacturers to retailers, and from government services to utilities; these customers are the major references listed in the annual reports and marketing brochures of all the major high-tech services providers.

As a consequence, all the major high-tech services vendors are very cautious and constantly check that their offers suit their customer's expectations when they sell and create delivery systems for professional services. As one vice president of Steria said: "We know that we have a quality solution, if the delivery comes smoothly. If there is a problem, even if it was not considered in the contract, you can be sure that the customer will not pay." Consequently, service providers rely on specific procedures for quality, not only in the presale phase but also in the post-sale phase (see Figure 3.6).

3.2.2.1 The quality procedure in the presale phase

In the high-tech services business, the selling is always easy (because you are selling promises); it is the delivery that is hard. Accordingly, successful high-tech services vendors move from one step of the presales phase to another through an explicit and documented procedure.

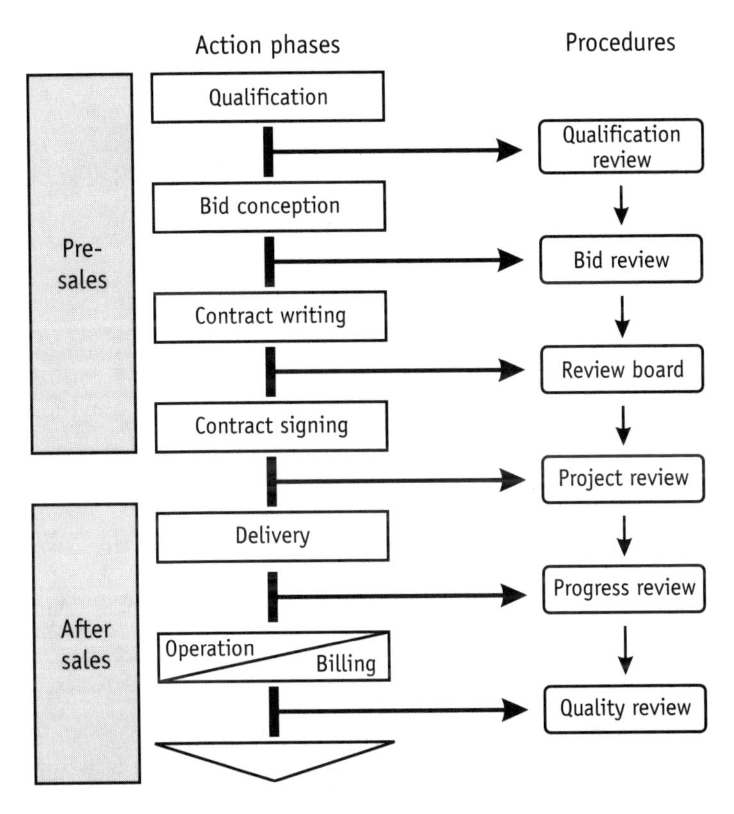

Figure 3.6 The procedures for quality of high-tech services for business customers.

Thus, each phase of the selling process, namely the qualification, the bidding, and the negotiation of the contract, includes specific reviews whose goals are to make sure that the vendor can build a solution that will effectively fit the customer's needs. If this is not the case, the vendor is better off quitting the project, at any stage, and explaining the reasons for doing so to the customer, thus keeping the level of trust unbroken. This is a must in the high-tech services business.

The qualification. Once a project has been identified, the first procedure for quality is the qualification review. It aims at clarifying all the issues to help the vendor to assess its strengths and weaknesses in answering customer needs. Most notably, it:

- Analyzes all the components of the service project:
  - The end-user needs;
  - The number and profiles of the decision makers;
  - The kind of third parties which may be involved;
  - The timing of the project;
  - The budget.
- Scrutinizes the context of the project:
  - The importance for customer relationships;
  - The competitor's position on the account;
  - The competitive impact of the project vis-à-vis the customer;
  - The significance of customer's value creation capacity;
- Makes a go or no-go decision.

If it is a go decision, then the operations during the sales cycle are detailed, the role of each member of the customer team is assigned, and action plans are made to prepare a bid.

The bid conception. The bid review is the second procedure for quality and involves the checking and the validation of all the elements of the proposal. Of particular note, the following are considered:

- The answer to customer's needs;
- The technical feasibility of the solution and its conformity to the in-house quality standards;
- The delays in building and delivering the solution;
- The compliance with legal constraints;
- The fit with the vendor strategy from a marketing and financial point of view (impact on sales, market share, image and profitability);
- The impact on competitive positioning in the market.

One must keep in mind that a proposal is not a contract. Although it is not legally binding, it represents a fair assessment of the commitments the vendor is ready to put into the contract. If this is not the case, it will certainly lead both parties into dispute later when the customer starts clarifying each assumption before writing the contract.

The bid review ends with a recommendation of corrections to be made within a given deadline for another bid review, or the proposal is approved for official presentation to the customer.

The negotiation of the contract. The third procedure for quality is the review board. It occurs once the customer has agreed on the technical content of the service and considers the content of the contract to be negotiated. The contract is to specify all the elements of the offer. If accepted and signed by the customer, it represents the official commitments of the vendor to deliver the service and may be challenged by law if the vendor does not keep up with its content.

The contract also covers many potentially contentious issues—like termination, late penalties, and financial liability in case of failure—and provides ways to address them. It includes technical information on the nature of the service to be provided, the date, and the way it will be delivered. It explains the mode of the service delivery and the mechanism of sign off as well as the customer's acceptance criteria.

The acceptance includes most notably the evaluation of the performance of the service delivered. In a solid contract, both parties agree on the timing of the test performing with live data as well as on the condition to settle any performance problem which could be found after the acceptance. Such an agreement also usually implies defining many variables to be measured and agreed upon.

The contract also fixes responsibility of both parties so that there is accountability. For many complex services, the corporate customer also has obligations. The customer has to make its demands clear and to sign off on the documented requirements before the development of the service starts. In many cases, the customer has to supply the test data to precisely state the acceptance criteria, and to carry out acceptance tests on the applications delivered without any delay.

A sound contract will also include specific clauses defining the conditions of evolution of the service with a change control procedure for any service upgrade, both during the development phase and after the acceptance of the delivery.

Intellectual property is also defined in the contract, as well as the level of transfer of knowledge which always takes place in any customized service development. Finally, a mechanism for termination is to be included in the contract defining the condition for termination as well as the payment during termination.

Therefore, the review board is extremely important because of the commercial, technical, financial, and competitive matters which are at stake. Its goal is to validate all the elements of the contract submitted to the customer. Of prime importance are:

- The legal and contractual terms and conditions of the contract;
- The delays, feasibility, penalties, and commitment toward the customer;
- The costs;
- The property clauses;
- The conditions for updating the solution;
- The billing conditions.

3.2.2.2 The quality procedure in the postsale phase

After the contract has been signed, the procedures for ensuring the quality of the solution start with the project launching review, which is followed by the progress reviews. The delivery phase is next, and, finally, the quality reviews are made during operations.

The project launching review. After the contract has been signed, the project launching review aims at transferring the prime contracting of the service application to the vendor. Again, it is an opportunity to check that the implementation of the application project will be in line with the customer's needs and expectations. During this review, the phasing of the project is detailed with a clear assignment of the responsibilities and commitments of both parties. People in charge are explicitly assigned to precise operations.

A plan is issued with project reviews arranged at regular periods of time or at any significant step in the operation before the official delivery. The project launching review also tries to evaluate all the risks associated with the phases of implementation and the possible slippage of time.

The delivery phase. The delivery phase takes place when the customer agrees to pay for the service that is performed and given over by the vendor. This acceptance of the service means that first it conforms to the conditions of the contract, but also that the customer is satisfied with the solution.

In large contracts for complex and sophisticated high-tech services, projects are divided into many subprojects of which each is submitted to a specific delivery phase. Consequently, there is not one, but many delivery phases, which are performed as milestones and allow the vendor to check that the whole project is still in phase with its customer's needs and expectations.

Operation. Finally, reviews are set up with the customer on a regular basis during operations to assess the quality of the service delivered and to discuss possible improvements. Furthermore, for major projects some companies like IBM Global Services have an outside consultant perform an independent and specific quality review at the end of the project in order to accumulate more experience.

3.3 Achieving quality of high-tech services for consumers on the Web

The mass market Web is the new prime business location where many new entrants to the market are fighting each other to become the most successful content provider. Offering a quality Web site is a primary way to build a strong competitive advantage, but the achievement of such a goal does not only require developing and offering quality solutions to customers. It also implies answering some broader needs of customers.

3.3.1 Developing a quality solution for the Web

To ensure a quality-oriented Web site and service implies thinking carefully about the information that is delivered to the customer in terms of content, but also in terms of the delivery media (for example, text, graphic, audio, or video) and design. The techniques for creating and channeling the information are also very important because they have a strong impact on the perceived quality of the Web service from the users' perspective because they are fishing for "relevant information in increasingly choked networks."

A good Web page design considers three areas:

- The *conceptual design,* which matches appropriate content formats with the different types of information needs of target audiences;
- The *navigation design,* which defines relations between pages;
- The *page design,* which gives a page a "look and feel" and navigation buttons or hyperlinks.

There are essentially many ways to develop a good Web site because of the numerous types of services offered on the Internet. However, here are some do's and don'ts for successful Web site design while avoiding common mistakes.

3.3.1.1 Conceptual design

1. *Escape overly long download times.* Traditional guidelines indicate 10 seconds as the maximum response time before users lose interest. On the Web, users have been trained to endure so much suffering that it may be acceptable to increase this limit to 15 seconds for a few pages.

   To achieve such results, one has to minimize bandwidth impact by keeping homepage topics to 500,000 bytes and pages to 25,000 bytes or less regardless of the audience. If pages exceed 25K, a distinct link has to be created to a low bandwidth version.

   Similarly, large (15K or more) image files have to be minimized. Many companies offer flashy screens full of graphics that take time to load, bring no real value to customers, and upset them because of the waiting time. Any long file must be checked beforehand with test customers.

Another way of saving time for customers is to keep the number of colors in an image to 256 or fewer.

2. *Shun gratuitous use of "bleeding-edge technology."* The mainstream users care more about useful content and good customer service than about use of the latest Web technology. However, some companies are using the latest technology before it is even out of beta test. This is the most effective way to discourage users because if their system crashes while visiting a site, many of them will never come back. Except for vendors who are in the business of selling Internet products or services, a sound practice is to wait until some experience has been gained with respect to the appropriate ways of using new techniques.

   To summarize, in the conception of a Web site, one must always give priority to efficiency and simplicity over showing off and complexity.

3.3.1.2 Navigation design

1. *Make it short and easy.* To avoid confusing the user, it is best to keep all navigation pages (the home page) to one screen. In this respect, it is important to keep the laptop user in mind whose computer screens have a 10.3-inch diagonal at most.

   To allow users to navigate sites easily, it is important to include "Back" and "Home" navigation buttons. Surprisingly enough, not all browsers have these buttons, making their navigation very difficult for the user and provides a very poor quality image, whatever the quality of the site's content.

   For similar reasons, it is better to index the home page topics because it helps the user to navigate more quickly and easily through the site and to go quickly to the information the user is seeking.

2. *Eliminate orphan pages.* Users may access pages directly without entering through the home page. Accordingly, the vendor must make sure that all pages include a clear indication of what Web site they belong to. Likewise, every page should have a link to the

home page as well as to give some indication of where it fits within the structure of the information space.

3. *Avoid any lack of navigation support.* A common mistake is to assume that users know as much about a site as the service vendor does. As a matter of fact, experience shows that users always have difficulty finding information, so they need as much support as possible in the form of a strong sense of structure and place.

   Consequently, the design of the navigation system must start with a clear understanding of the structure of the information space. Then this structure has to be communicated very explicitly to the user, using as much support as possible to let users know where they are and where they can go. A site map and good search software are always of help for the first-time user. Once the user knows the site well, he or she will directly access the needed information.

3.3.1.3 Page design

1. *Keep it short.* The length of individual pages should be no more than four screens long. An exception is when it is more convenient for an individual to save a single file rather than multiple files such as an article or instruction. Even in these cases, a lengthy document should be broken into separate files giving viewers the option of opening the entire document or specific parts of the document.

   Similarly, it is advisable not to design long scrolling pages, mainly because only about 10% of users scroll beyond the information that is visible on the screen when a page comes up. All critical content and navigation options should be on the top part of the page.

2. *Test all graphics early and often on something other than the computer you are working on.* Test on both graphic and text browsers. Your page may have great images that turn to ugly glop when applied to different computers.

3. *Make sure that the images add value.* A string of buttons in an image map does not always add value. However, images that show the relationship between the different items do add value.

4. *Every page should contain:*

   - Distinctive file names so viewers can easily download pages.
   - The name and e-mail address of the data maintainer(s).
   - The "address" tag at the bottom of every page. This tag identifies when the page was last updated.
   - A navigational link or button that returns the viewer to the home page.
   - Any known omissions or problems with the data.
   - A description, copyright notice, origin, and authority of the data (if appropriate).
   - The frequency with which data is updated or expiration date (if appropriate).

5. *Never include page elements that move incessantly.* Many Web sites tend to have too much blinking and constantly running animation. This is a mistake because moving images have an overpowering effect on human peripheral vision and actually prevent users from being able to read the text. Consequently, a good Web page gives the user some peace and quiet to effectively see and understand its content.

6. *Respect standard link colors.* The ability to understand what links have been followed is one of the few navigational aides that is standard in most Web browsers: links to pages that have not been seen by the user are blue, while links to previously seen pages are purple or red.

   It is important to be consistent with this standard and not to use other colors. Otherwise, it can mislead customers who get lost in the site and may become frustrated with the service.

7. *Update information.* Users are looking for the right information and performance at the right time. They do not want to have outdated data, however some older information may still be of use. As a consequence, the maintenance of a site is a key element of success for a service vendor, although very often most of the vendor's people would rather spend their time creating new content than maintaining existing content.

   In practice, maintenance is a cheap way of enhancing the content on a Web site since many old pages keep their relevance and should be linked into the new pages. However, some pages are better off being removed completely from the server after their expiration date.

3.3.2 Answering broader consumer needs

As of today it is estimated that more than half of Web users are not likely to make on-line purchases in the future because they are not happy with the quality of the service provided by the Web "per se." Many individuals are worried about its lack of security, its absence of law, and its attack on privacy.

In this case, the quality of an Internet service must be extended not only to its content and its design but it must also answer the broader issues of security and privacy that are central to the customer's evaluation.

3.3.2.1 Answering the lack of security and billing issues

The nature of the Internet clearly increases transaction risks because of the lack of security and legal recourse. As a consequence, the more expensive a product or service, the less likely it will be purchased on line. This is because, all other things being equal, the degree of reluctance to purchase a solution on the Internet is a function of the level of potential loss.

In general, consumers will seek to minimize their risk when purchasing high-ticket items by seeking more prepurchase information and minimizing transaction risks. Answering these two expectations is of prime importance for offering a solution that will be perceived as a quality service by consumers.

Commerce on the Internet has suffered from the lack of readily available and appropriate payment mechanisms. Today, payment is generally made via credit card but concerns over security have lead many users to rely on fax or telephone for authorization details.

In the United States, where an array of automated systems are available, most businesses continue to bill their customers with paper invoices and make payments using paper checks. The Federal Reserve estimates that in 1993, only 3.8% of business-to-business payments by transaction volume were made electronically using either the Clearing House Interbank Payments System (CHIPS), Fedwire, or Automated Clearing House (ACH) transfers.

Security is one of the major impediments to the growth of Internet sales. The fear that hackers will be able to intercept unencrypted credit cards and consumer details transmitted over the Internet has restricted sales growth, especially for high value products. Various proprietary systems have been designed to protect on-line transactions, but no standard has yet emerged.

The latest technology enables credit card information to be protected with Secure Sockets Layer (SSL). To secure electronic downloads, two separate Internet providers may transmit the software itself and the password. Once the password is entered and received by the software, the package self-destructs.

However, O'Reilly & Associates, an Internet research, software, and publishing company, recently conducted a security census of 648,613 public Web sites using Netcraft's Web query technology. It claimed to have found that only 10% of them offered the SSL protocol, which the study cited as one of two criteria for considering a Web site "fully enabled" for electronic commerce. Furthermore, just 5% of those sites—half a percent of the total sites—offered *third-party certificates*, a second criterion for electronic commerce, according to the study.

However, according to the study, 500 sites per month had authentication added through trusted third-party certificates since the beginning of 1998. This number is up sharply from the 200 to 300 sites per month that had them added during the last quarter of the previous year.

One should note that the majority of sites studied were in the United States. In fact, the United States hosts 70% of the SSL sites and 77% of the sites with valid certificates while approximately 60% of the fully enabled commerce sites also offer strong encryption.

3.3.2.2 Legal recourse issues

Security is not the only risk that consumers face when purchasing products on the Internet. Vendors can also renege on their agreements. For example, there is the financial risk that the product will not arrive after payment has been authorized resulting in a direct financial loss for the purchaser.

There is also the functional risk that the product will not operate as advertised by the vendor because there are currently no agreements regulating electronic commerce between countries. If a consumer in the United Kingdom orders a product from a site operating in the United States, there is little legal recourse if the product fails to arrive or malfunctions after arrival.

Determining the legal jurisdiction within which to pursue litigation claims may be problematic in itself. For example, a product ordered from a site operating in the United States may be shipped from Singapore by a company incorporated in Switzerland. In which jurisdiction should a claim be pursued—the United Kingdom, the United States, Singapore, or Switzerland?

A number of tools can be used to mitigate this risk, and they offer exceptions to the general rule that highly priced items will not sell on the Internet.

Brand names and reputation play an important part in reducing the risk of reneging on contracts. Companies with a large volume of sales across many channels cannot afford to renege on Internet sales for fear of damaging their reputation in other markets.

In addition, high levels of brand advertising signal commitment. Therefore, large reputable companies with a strong brand name could be expected to sell more on the Internet. The large multinational firms are also assisted by having branch offices where customer complaints and problems can be resolved and goods can be exchanged face-to-face.

Also potentially mitigating consumer risk are warranties and guarantees, but only to the extent that they are enforceable. Again, the existence of a strong brand name or reputation signals that customer assurances will be honored.

Recently, companies with strong brand identities, such as Microsoft, Time-Warner, and AT&T, have established virtual malls on the Internet. For smaller companies, establishing an Internet presence in a virtual mall reduces consumer risk because it effectively links the reputation of the vendor with that of the mall operator. It is in the interest of the mall owner to police the actions of its members or risk damaging its own reputation. Malls also provide the additional benefit of reducing the amount of search required by consumers because everything is metaphorically "under one roof."

3.3.2.3 Answering privacy issues

Privacy is an issue difficult to size because what is private changes from one person to another as well as from one culture to another. But in January 1998, the U. S. Federal Trade Commission (FTC) concluded that the growth of the commercial side of the Internet inevitably generates more personal information and that privacy policies should be developed to give users proper notice, choices, and security. Consumers in the United States are particularly concerned over privacy issues where businesses and computers come together.

A 1995 Harris Survey said 82% of Americans are concerned about threats to their own privacy and 78% believe that consumers have lost control over how businesses use personal information. The Internet community appears hypersensitive to such issues. Consumers who use the Internet are likely to be better educated and less willing to give out information, unless they trust the recipient.

Consequently, service providers on the Internet need to clearly explain to users what information they collect, how they collect it, and what options users have to quit. Companies must be absolutely clear about the value and intended use of information. The more users know about what actually happens with the information, the less likely they are to speculate about abuses.

In the case of cookies, for example, Netscape users are now far more accepting of cookies today. Initially the idea that a Web site could write and read information to a user's hard disk raised concerns that were easily distorted. The more people understand how restrictive this capability really is, the more comfortable they are with the notion.

A number of new capabilities are being developed to give users greater control. For instance, Navigator 4 will enable users to permanently deny cookies, so that they do not have to respond to the server's query each time a cookie is presented. Utilities are also available to automatically delete cookie files at the end of each session. This permits a user to benefit from session keys, while eradicating persistent cookies.

A site that collects very detailed information about what users do on its site may, for instance, never attach a real name, phone number, or e-mail address to the data but simply use the cookie identifier.

In concluding this section about the issues of Internet privacy and security, one must note that many millions of people trust others with their personal financial information in the context of normal business activities. Examples include ordering over the telephone, passing a credit card to an unknown waiter, and even signing direct debit mandates. If an error occurs in these types of transactions, customers trust the service provider to correct the error. So far, the Internet is expected to support a level of trust and security which is not observed in everyday life.

Actually, billing safety techniques are now available and trust relationships can be established in electronically mediated discussions where it becomes easier for an individual (or group of individuals) to seek retribution against those that break the rules within an electronic community. Evidence of this can be found in the tendency to attack those that try to advertise in academic discussion groups (mail bombs) for instance.

With time, education, and quality solutions, one may safely bet that e-money will rise and bring about the most fundamental changes in trade since the invention of paper money.

3.4 Chapter summary

Besides customer focus, quality is the second lever for achieving success in any high-tech services business. Quality can be defined as a contractual commitment that sets up the various levels of service and performance provided to the customer.

This definition is very similar to the one given in product manufacturing—delivering the right product to the right customer—with a major difference. Indeed, actions to improve product quality have concentrated mostly on the removal of product failures. Likewise, a significant part of the service quality relies on the elimination of defects before the customer sees the service, and by focusing on the design and implementation of the service.

However, this is not enough because of the essence of high-tech services defined in Chapter 1. To ensure the repeated quality of the performance, service quality has to be an ongoing component of the service's operations. Consequently, high-tech services vendors are developing and using a variety of specific measurement tools to constantly monitor their performance versus their customers' expectations and perceptions. As a result, service quality is not achieved in the same way depending on the nature of the customers, being businesses or consumers.

Failures in service quality are not limited to the technical problems impeding the due performance of a given service. A poor image of quality for a high-tech service comes more generally from the gap between customers' expectations of a service and their perception of the service provided. Quality oriented high-tech services firms intend to close or decrease such a gap. In order to do so, they must also lessen:

- The difference between the user's expectations and the specifications of the service;
- The variance between the specifications of the service and the actual service delivery;
- The gap between the service delivery and the customers' perception of it;
- The gap between the customers' expectation of a service and their perception of the service provided.

To ensure that they are offering top quality solutions to their corporate customers, high-tech services providers first rely extensively on the use of state-of-the-art technology, which allows for maximum performance but is not without risk. Consequently, achieving the quality for those services requires a careful software development process using strict architectures and the automation of repetitive and low value operations.

It also demands active risk management—using models such as the CMM or SPICE—to identify the risks, to anticipate their potential impact, to make contingency plans, to follow up throughout the project, and to wrap it up in order to build experience.

The last step for technical quality is the willingness to constantly improve performance, which calls for a pool of experts, the establishment of transmitters to ensure that these methods are used in the day-to-day activities of software development units, and perseverance and endurance.

Nevertheless, technical quality is not enough to ensure service quality. All the major high-tech services vendors constantly check that their offers suit their customers' expectations when they sell and create the delivery system for professional services. They know that they have a quality solution if the delivery comes smoothly. Consequently, service providers rely on specific procedures for quality not only before but also after the sale.

Each phase of the selling process, namely the qualification, the bidding, and the negotiation of the contract, includes specific reviews whose goals are to make sure that the vendor can build a solution that will effectively fit the customer's needs. After the contract has been signed, various reviews are performed during the implementation of the delivery system, which is followed by the delivery phase, and finally, the quality reviews, which are done during operations.

Regarding Web-based high-tech services for consumers, offering a quality Web site is a primary way to build a strong competitive advantage. This implies putting a lot of work and creativity in the conceptual design, the navigation design, and the page design of the Web service.

The achievement of a quality Web site also implies answering some customers' broader needs. Many individuals are still worried about the lack of security on the Internet, its absence of law, and its attack on privacy. Consequently, the quality of an Internet service must also answer broader issues of security (through technical solutions, guarantees, and a strong brand image) and privacy (thanks to an effective communication) which are central to the evaluation by consumers.

References

[1] Parasuraman, A., V. A. Zeithaml, and L. L. Berry, "A conceptual model of service quality and its implication for future research," *Journal of Marketing*, Vol. 49, Fall 1985, pp. 41–50.

[2] McKenna, R., *Real Time*, Boston: Harvard Business School Press, 1997.

[3] Cooper Ramo, J.C., "How AOL Lost the Battles But Won the War," *Time Magazine,* September 22, 1997, pp. 72–75.

[4] Ketani, N., *De Merise à UML*, Paris: Edition Eyrolles, 1997.

4

Organizing Human Resources for High-Tech Services

As we have seen in Chapter 1, high-tech services can be defined as a mix of information technology, methodology, and personnel, which aims at bringing a valuable solution to customers. This means that the highly knowledgeable people of high-tech services firms have a significant impact on the solution delivered, either through their industry expertise, or through their command of technology. One executive of Andersen Consulting declared: "Our skills are our people and our assets are their knowledge in conception, development, and implementation of solutions for our customers."

Why are the people so important? This is what we are going to see now before considering the ways high-tech services companies leverage their human resources through specific organization and sound management.

4.1 Why are human resources determinant in the high-tech services business?

It does not come as a surprise that personnel is the third key success factor in the high-tech services business for three main reasons. First, employees, such as consultants or maintenance technicians, may have a very significant (positive or negative) impact on customers because they are meeting them face-to-face. Secondly, the experience, the motivation, and the knowledge of employees, such as software engineers or developers, are determinant in the quality of the solution offered. Thirdly, the costs of people are so important in this industry that they can make or break the profitability of a firm.

Let us now consider these three points in a more detailed manner. However, before starting, one needs to differentiate among the various categories of employees within a high-tech services company, because they cannot be considered as a homogeneous lot. Indeed, they have significant differences of importance and status according to their relative contribution in the service value as well as their relative exposure to the customers.

It is a fact that, as shown in Figure 4.1, all high-tech services are not always people intensive. Some services, like consulting or system integration, require a significant number of personnel, while maintenance or outsourcing services require equal amounts of people and technology. Finally, network services rely much more on technology than people: once the network application has been designed and developed, it does not need a group of people to keep it running.

Besides, employees interface differently according to the kind of service offered. Some employees meet the customers face-to-face (or on the telephone) very often, such as consultants performing strategic consulting or maintenance technicians performing their job at a customer's location.

Other employees meet customers less frequently, such as software engineers working on systems integration or application software, who may spend a large amount of time working in house on the design and development of the solution. Regarding outsourcing services, most of the people performing the service are invisible to the customer, likewise for the people running network services.

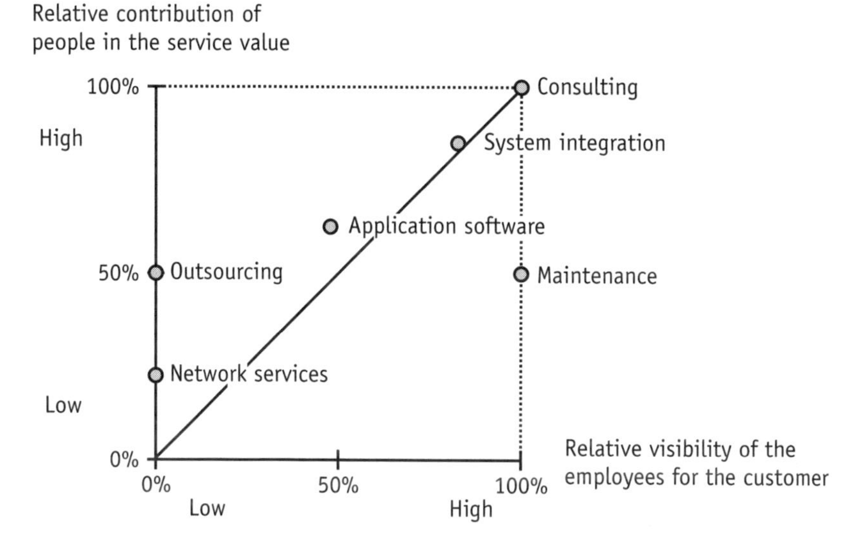

Figure 4.1 The relative importance of employees in high-tech services.

Another important distinction relates to the profile of the various employees within a high-tech services firm. They vary according to the type of service offered. Figure 4.2 gives a raw list of the main categories of people performing a given service, outside of the sales and marketing people.

As we will see later when studying the management of people, one cannot really compare a consultant with a software engineer or a maintenance technician. Their profiles, their knowledge, and their expectations are different and need to be considered separately.

4.1.1 The impact of employees on customers

Leading service organizations are quite aware that their people make the difference at the customer level because they are very often part of the performance experienced by the customer. This is of prime importance for all high-tech services where employees are building most of the value of the services offered.

The value of a consulting service depends almost exclusively on the analytical capabilities and the presentation capacities of the consultant (or the team of consultants) who is working with a customer and has an

Type of services \ Type of customers	Businesses	Consumers
Strategic consulting	Consultants	
System engineering	Software and hardware engineers	
System integration	Software and hardware engineers	
Support	Technicians	Technicians
Outsourcing	Consultants, software engineers, and technicians	
Network services:		
• On-line information	Software engineers, application engineers, and designers	
• E-transaction	Software engineers, application engineers, and designers	
• E-business	Software engineers, application engineers, and designers	

Figure 4.2 A list of the different types of employees involved in high-tech services offered to customers.

intimate knowledge of the customer's business and the challenges to meet. When good consultants leave one company for another, it is not infrequent to see customers follow, switching to the consultants' new firm.

Similarly, in some cases only a handful of highly skilled and experienced software engineers are able to manage the integration of complex and sophisticated systems involving intricate programs and numerous pieces of hardware on a specific project for a large dedicated customer.

Having these specialists move to another company can endanger not only the achievement of the solution but also the credibility of the company vis-á-vis its customers. Should the project be significantly endangered, there is no doubt that the customer will go to the new company where these specialists have moved.

To a lesser extent and for similar reasons, the resignation of highly skilled and experienced application software specialists can have dramatic consequences for a high-tech services firm.

Indeed, these companies are not providing standard solutions to their corporate customers; a good application software must have both a strong software design competence but also a far-reaching aptitude to conceive a solution adapted to the specific needs of a given customer.

4.1.2 The impact of employees on quality

We have shown in Chapter 3 that it is important to differentiate between the "perceived quality" as grasped by the customer and the "intrinsic" quality of the high-tech services delivered. One of the direct impacts of contact employees is on the perceived quality [1]. The more credible the employees, the more comfortable the customer feels about the performance and the experience of the service, that is, the perceived quality of the service.

A typical example is when a question is asked or when a problem occurs. What will make or break the perceived quality of the service is the ability of contact employees to answer questions quickly and correctly, or their proficiency at correcting the problem or explaining its causes, while presenting a timed action plan to fix it.

This is only the tip of the iceberg. Before even thinking about the perceived quality, a high-tech services firm must have people who are able to build a quality solution per se. We have shown that this implies relying both on technical tools and procedures.

Considering that most of the technology (hardware and software) used in building such a solution is less than three years old, it is a tough challenge to find people who have a good command of it. As one executive lamented: "Everything is moving too fast in our business. Before, programs used to be written in Fortran, then in the so-called fourth generation languages; now with the Object-Oriented Development software, we cannot find all the software specialists we need, and we cannot use older ones who have not caught up with this new programming technique."

Procedures for quality, both in the presale and postsale phases, also require finding outstanding people who must have:

- Discipline, because each step of the procedure demands a lot of work in preparing for various meetings, then documenting the actions to be taken, then implementing them before moving on to the next step;

- Good communication skills to be able to listen to customers and then talk to them at the various stages of the project;
- Strong teamwork inclinations, because the building of a high-tech service calls for interfacing with numerous specialists inside and outside the firm.

This profile does not fit exactly with many skilled technical employees, who are more comfortable writing software code in the office on a desktop computer than going to a customer's location and making a presentation.

As a consequence, people with the capacities to use the technical tools and conform to the procedure while enhancing the perceived value of a high-tech-based service to customers are a rare breed and must first be identified within the company and then carefully nurtured.

4.1.3 The impact of the costs of people on profitability

The last reason why people are so important in the high-tech services business is the sheer influence they have on profitability. Figure 4.1 illustrates the typical cost structure of a high-tech services firm. Cost figures may vary from company to company. Actually, the operational margin in the professional services industry in 1997 was about 11%, slightly higher than in the case of the example.

In any case, as one can see, employees are eating the lion's share of the revenues, representing up to more than two-thirds of the costs as in the example of the company shown in Figure 4.3. This amount echoes the contribution of the employees to the value delivered to the customers.

It may come as a surprise for people not familiar with high-tech services that the cost of people is so heavy in a business where technology (and its associated costs) is so significant. On the one hand, it is important to remember that the costs of information technologies have been constantly going down for the past 20 years, while the performances have been increasing dramatically.

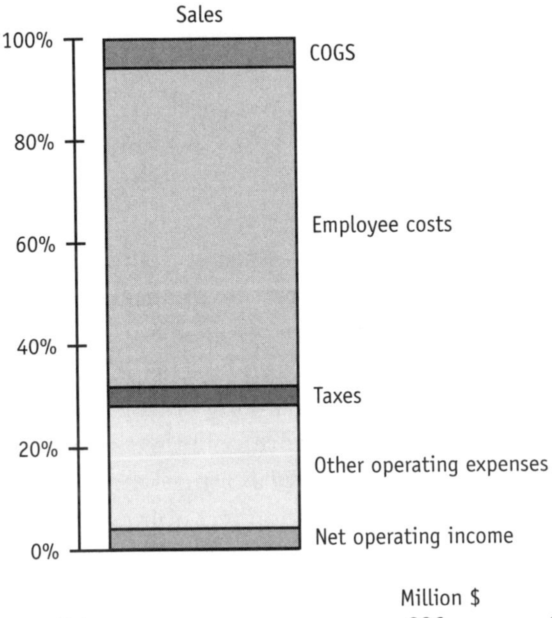

	Million $	%
Sales	**206**	**100%**
COGS	10	5%
Employee costs	**136**	**66%**
Taxes	7	3%
Other operating expenses	46	23%
Net operating income	**8**	**4%**
Financial expenses	2	1%
Exceptional expenses	1	1%
Corporate income tax	2	1%
Amortization of goodwill	1	0%
Net profit	**2**	**1%**

Figure 4.3 The weight of the employees' costs in the cost structure and their impact on profitability of high-tech services. (*Source:* Steria, Annual report, 1997.)

On the other hand, this example illustrates the fact that, more than technology, it is the people who are using the technology that are the core competence of high-tech services firms. These people usually have a high level of training and expertise, and are currently in high demand.

In Germany, starting salaries at Debis or SAP are between $35,000 and $45,000 for a graduate in computer science; this is between 50% and 100% more than the entry-level salary for a chemical engineer. In the United Kingdom, according to *Computing*, a British trade weekly, experienced programmers with management skills may earn $250,000 and more.

Actually, the high-tech services cost structure is not very different from those in many services firms whose key characteristic is to be labor intensive.

For managers, the consequences are straightforward. To achieve profitability and success requires careful management of the human resources in the company. Employing too many people may overload operating costs, while employing too few may impede the growth of revenues, leading in both cases to a sinking bottom line.

The problem is getting even more complex because of the erratic nature of the information technology labor market: after the recession in the beginning of the 1990s, information technology specialists could not find work, so many students moved into other fields leading to a "missing creation of programmers and specialists," according to one executive at Experian. Even today, about 40% of computer science and electronic engineering positions inside European universities are unfulfilled. In Germany, only 60% of the 11,000 software engineering student places are filled.

Today, with the growth of high-tech services markets, there is a skill gap throughout western countries. In Europe, it is estimated that more than 200,000 jobs were unfilled in 1998 because of a lack of available people with programming skills and project management proficiencies. Many executives share the complaint of the CEO of Debis, Klaus Mangold, when he said: "We could take on more projects if we could hire more people." His American competitors feel the same. For instance, Microsoft recruits a high percentage of programmers in Asia.

Even the difficulty of finding a knowledgeable and trained workforce may have consequences on the bottom line. In 1997, when the U.K. software consulting firm, Logica, disclosed its difficulty in finding skilled information technology personnel, its share price dropped by $3!

As one executive stated: "If you want to win in this business, you always need to have the right number of people with the right competencies at the right time." This probably could be said for many other businesses, but such a goal is both a necessity and a challenge in the high-tech services business because of the complexity of the business problems that need to be tackled, the speediness of change in technology, and the high demands of customers combined with the sheer impact of people on profitability.

To achieve such a goal, successful high-tech services firms constantly monitor their people and performance; they also try to anticipate their needs, both in quality and quantity; and, they seek to keep their people motivated. To do so, they rely both on an efficient organization, as well as on sound management of their employees [2].

4.2 Organizing the human resources of high-tech services firms for performance

The most successful companies in the high-tech services business have all spent a significant amount of time designing their organizations in order to leverage their human assets. Nevertheless, they are constantly fine tuning it because of the swift evolution of their customers, their environment, and their own employees. The key word is *flexibility* and adaptation around two or three prominent dimensions that provide the framework of the organization.

High-tech services companies are not heavily weighted oil tankers that need seven kilometers to turn around. They are much more like the new breed of hybrid boats that are powered by engines, but also able to sail when the wind starts blowing, and have a long-range capacity, but also the ability to maneuver as fast as a small yacht.

Before analyzing the mode of organization of the major high-tech services firms, we must take note that this organization is significantly different from product-oriented firms. The reason for this is that the "adding value" process of high-tech services vendors are very distinct and peculiar and have some idiosyncratic characteristics embedded in them.

4.2.1 Organizing for business-to-business high-tech services

Many executives in the high-tech services business would agree with the comment made by a vice president at Compaq: "In our business, the value chain starts from understanding the market to creating and producing a solution, then to selling the solution to the customer with reciprocal commitments, then to delivering the solution, then billing and ultimately receiving payment." Such a model is different from the one used to bring value to customers through a high-tech product. Indeed it drives many consequences as far as the organization of a high-tech services firm is concerned.

Most high-tech services firms are structured as presented in Figure 4.4. There are three main functions (marketing, engineering, and operations) and two major support functions (R&D and human resources).

Marketing and sales are in charge of identifying the needs of customers, negotiating the project contract, and assessing the customers' satisfaction with the perceived quality. The engineering function is in charge of the conception and design of the high-tech services. The operations function is in charge of running the service. The maintenance function is not identified formally because most of its activities are in house; it belongs to engineering or operations according to the nature of the service offered.

One of the consequences of such an organization is that there are three different categories of technical population: the architects and engineering specialists, the operations specialists (for instance, systems engineers), and the customer care specialists, who are between the customer and operations.

The R&D function is very limited in size. In many companies, such as Compaq for instance, it is very much in touch with operations (also called "field" or "delivery" teams) because as one executive stated: "We need to develop more and more methodologies and tools (such as application software) both to differentiate ourselves from the competition and to always make our customers feel confident about our delivery of

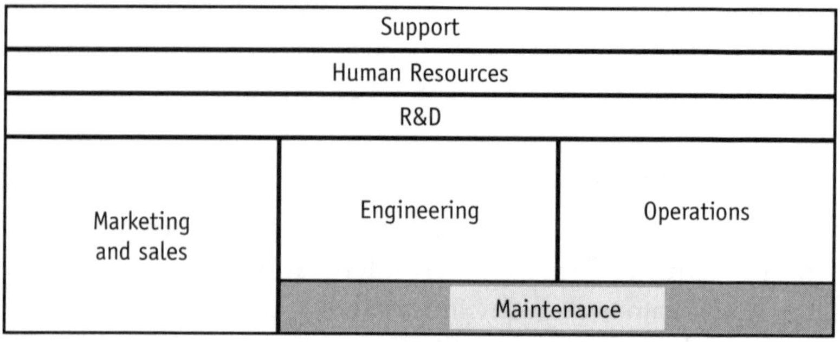

Support
Human Resources
R&D
Marketing and sales
Engineering
Operations
Maintenance

Figure 4.4 The value chain of high-tech services for business customers.

information and applications at the exact time required." Such emphasis on the interaction between the R&D department and operations is also the rule in the software and service group at Hewlett-Packard.

The human resources function cares about the employees, as we will see in a detailed manner in Chapter 5. Other support functions include the legal department, which looks at legal issues, the finance department, which must work closely with marketing and sales to set up the price of an offer, and the department in charge of quality, which is constantly in touch with engineering and monitors operations on a regular basis.

This specific way of delivering value to corporate customers translates directly in the organizational chart. Most of the successful high-tech services vendors are organized along a supply/demand matrix with a geographical dimension added when those companies are global (see Figure 4.5).

As one executive of Cap Gemini explained: "We are organized along three axes:

- A market driven axis, according to the various kinds of accounts we want to serve;
- A technical axis because each service does not require the same competence. Strategic consulting does not rely on the same skills as information technology project management. Similarly, one

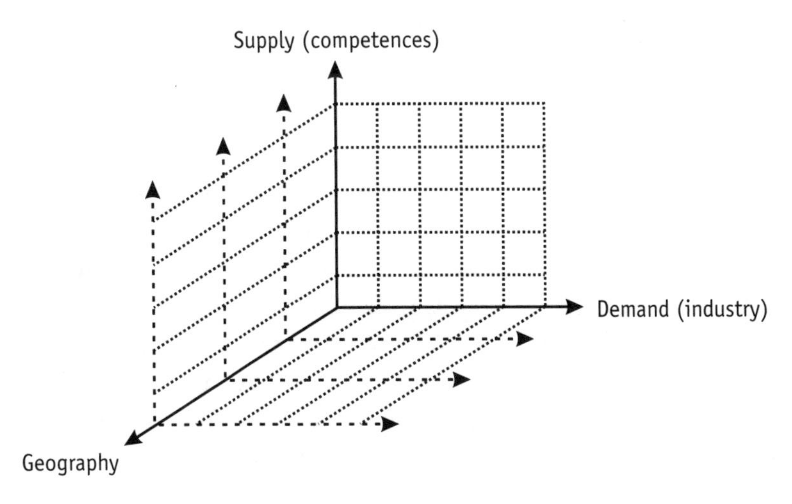

Figure 4.5 The organization of successful high-tech services vendors.

works differently on a typical mainframe or on Unix based computers.

- A geographical axis because we are in a business where the local dimension is important: the design of an information system is different whether it is to be used in Finland or in Spain."

The only difference in organization among leading high-tech services providers is the geographic dimension. Some companies are considering their markets at the local level. As one executive of IBM Global Services said: "Our organizational chart combines:

- Decentralized operational units, with P&L (profit and loss) responsibilities, specialized by industry and country (though we can provide one single interface if a customer requests it).
- With centralized transverse units, in charge of the competencies, strategic marketing, and quality, to enhance at a global level the coherence of our brand image, our offers and the associated quality, as well as the transfer of experience and the flow of our resources."

The executive president of Syseca, a French software service firm, confirmed: "You must definitely be decentralized to be as close as possible to customers, with a single methodology and a very proactive management of your brand image." The service divisions of Compaq and Hewlett-Packard are organized along the same principles.

However some companies like Andersen Consulting have dropped the geographical dimension. "We are organized on a worldwide basis along a supply/demand matrix. P&L are by industries, while resources management is by competence. In our current organization, the national dimension is gone. The French subsidiary no longer has its own P&L," noticed one French executive of Andersen Consulting.

Such an organization helps to focus more easily on global customers. Its down side may be to lose a local sensitivity to less international customers. In any case, the question is much more connected with the strategic option of globalization than with the very nature of the high-tech services business.

What is important to remember is that successful high-tech services vendors try to constantly manage the balance between being customer focused and being production oriented. Through their organization they manage to answer the variety of needs of their different corporate customers while accumulating knowledge and building a standard core of competence which can be formalized, repeated, and perhaps automated. Such a process of accumulating knowledge allows them to decrease their costs as they go down the experience curve and achieve economies of scale in a similar way as companies that are manufacturing products.

However, it is important to notice that not all high-tech services can easily be industrialized as shown in Figure 4.6. Most of the network services, maintenance services, applications services (at least for the core systems), or some outsourcing services (such as facilities management or call center management) can be industrialized through automation and repetition. Other services, such as systems integration or consulting, vary heavily according to each customer's situation; consequently, they are less easy to standardize and replicate from one customer to another.

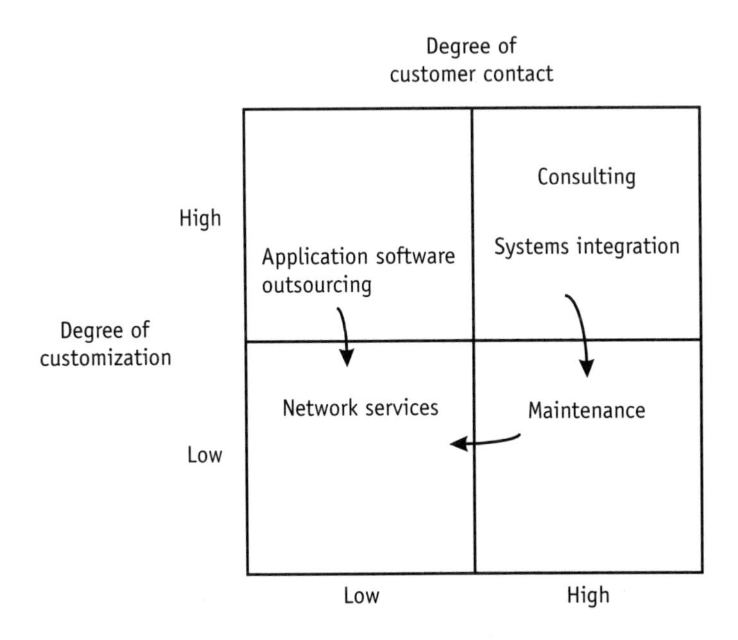

Figure 4.6 All high-tech services cannot be easily industrialized.

Another key element to bear in mind is that successful high-tech services firms are putting in place cross-functional teams as a key success factor to building customers' satisfaction (and competitive advantage) through an effective combination of all their various resources.

In order to offer a positive and significant experience to customers, they constantly strive to break down the so-called "silos" or "chimneys." It is a well-known fact in large companies that big units or functions tend to live on their own. They are called silos or chimneys because they stand alone and communicate at the top, and often only to compete and struggle with one another to get the maximum resources. Behaving in such a way is proscribed when making high-tech products because of their short life cycle, which does not allow any room or time for internal infighting, unless companies wish to lose ground quickly to competitors.

In the high-tech services business, working in silos would mean that valuable expertise is kept inside the function and is not shared with the others. A company operating in this way is not able to identify the complex needs of customers and to fulfill those needs given the sophistication and intricacy of high-tech services. Indeed the service solution subtly combines business knowledge, strategic expertise, innovative technology, virtuosity, and operational know-how. Such blending can be achieved only by teaming the best people at the right moment.

Consequently, cross-functional teams are the standard way of doing business at all the major firms we have interviewed. Marketing, engineering, and sales people team together at the presale phase. Engineering, operations, and quality operate together at the postsale phase. R&D, marketing, and engineering work jointly when trying to figure out new potential markets and solutions.

Furthermore, some firms are performing cross-functional evaluations on a regular basis. At Andersen Consulting, for instance, every three months, partners do a job review on a job outside their current function; such a procedure brings an external viewpoint to the team, as well as gives partners the opportunity to expand their own knowledge.

An engineer, a consultant, a technician, and a lawyer will see the same event with different spectacles and have different understandings. Consequently, through the diversity of the profiles of the team members, people in cross-functional teams get a broader and richer vision of the project or the job they are working on. Experience also shows that people in cross-functional teams tend to concentrate on finding a solution across departments instead of finger pointing.

Finally, the common sharing of knowledge makes everyone more open to change and willing to try innovative solutions to meet customers' requirements and business objectives. Successful high-tech services companies are truly "learning organizations," which as one executive of Andersen Consulting stated, "are always reinventing their business."

4.2.2 Organizing for business-to-consumer high-tech services

Reaching digital consumers requires a specific organization, should the service vendor be a specialized firm or just adding an on-line activity to a more traditional business, such as in the case of banks, newspapers, TV networks, and so on. Practice shows that in most cases these vendors go through a three-stage process as far as the organization is concerned [3]. They start by doing some experimenting and testing; if this goes well, they move to developing the service; if this is successful, they make it an "ongoing" service (see Table 4.1).

Table 4.1

The Stages in the Consumer On-Line Organization

Stage	Experiment	Development	Ongoing
Goal	Being there	Setting up business Delivering value	Making profit
Activities	Creating an on-line presence Gathering internal interest and skills	Developing business and technical expertise	Working closely with the other departments
Structure	Informal	Emerging	Integrated
People	Few people, working part time	5–20 people working full time + limited external competencies	More than 20 full time employees + extensive external competencies
Skills	Few experts with little proficiency	In-house experts coming out	Articulated and specialized experts

The objective of the experiment phase is to be there and explore the new opportunities the Web—or other technological innovations—can provide. In many cases, initiative starts with individuals, either technically oriented engineers or creative marketers who, for instance, want to set up and maintain a Web site. They may have little Internet proficiency, but enthusiasm makes up for this lack of expertise. These people work part time (sometimes on their own time) and their efforts are not integrated in a formal structure. Typically their first Web site is a collection of standard marketing pamphlets with few adaptations, and it does not take a very long time for these people to figure out that customers are looking for something more detailed and interactive.

Imperfect as it may be, the first Web site may create interest within the company. Soon such projects may get the close attention of a senior executive, who will allocate limited resources to a few people, often working part time on the project, to assess the business pertinence of going digital.

This acknowledgment is a must for the informal group to gain in-house credibility. Without top management support, it will remain one experiment among many others in a large company.

Once the organization endorses the ongoing experiment, guidelines are made to determine which activity can go on line with which brand and who can start the process. Such decisions are made at the senior level within cross-functional committees, which include representatives from the marketing, information technology, and legal departments.

If the experiment goes well, the next step is to develop the on-line service activity as a business that delivers an added value to the consumers. To do so, the priorities are to build business as well as technical expertise.

In this second phase, the group gets a formal manager who usually reports to the information technology department (but has one senior executive overseeing his or her activities). It expands to 20 full staff people and starts differentiating its marketing activities from its technology-based activities.

On-line technical and marketing expertise are built inside through training and experiments, as well as bought outside from a limited number of external sources like advertising agencies, or Internet specialists. This is the time when extensive work is done to (re)design all the material that is put on the Web so that it brings real value and satisfaction to the service users, according to the principles we have seen in Chapters 2 and 3.

If everything goes well and if the on-line services can prove that they are adding value to the overall business of a company, the ultimate step is to make it an ongoing business and to establish its plain dimension within the organization besides traditional information systems.

In some companies, the on-line services activity is no longer a cost center and its manager has a profitability goal to make. To do so requires working closely with other departments and creating a formal connection with other companies' core business. The structure of the on-line service department is fully integrated. It may have more than 20 full time employees from operations, engineering, and marketing, mirroring the typical business-to-business high-tech services vendor as shown in Figure 4.4. It may also be managing a whole array of external partnerships to get the expert knowledge required to get first class achievement.

Such an evolution does not come easily. In many cases, the new on-line service organization has to struggle with traditional functions at best, because it is changing their habits of doing business, and at worst, because they fear losing power to the newcomer. Likewise, any change in the organization of consumer high-tech services must be led and backed by top management. If not, it is doomed to fail.

Electronic service business success does not come simply from choosing the right technology; it also requires fundamental changes in organizations, corporate behavior, and business thinking. "We restructured our entire company around the idea of connectivity," said Rob Rodin, CEO and president of Marshall Industries Inc., a $1.3 billion electronic components distributor in El Monte, California. "The Internet is not the end-all, but connectivity is." Connectivity at Marshall means 24-hour call centers, 24-hour live help via interactive chat sessions on its Web site, and on-line supply-chain management among its buying and selling customers. Among a host of internal organizational changes paving the way, Marshall scrapped all sales commissions and management incentives based on specific objectives in favor of one reward criterion: overall corporate profits. The idea was to prevent friction between its traditional sales processes and the new sales channel of cyberspace.

Companies that want to achieve success in the ever-changing world of the Internet must understand that extranets and Web links are not only a new sales channel, but also a fundamental challenge to existing structures and relationships.

This is the reason why high-tech services vendors do not rely only on an efficient organization, but also put a strong emphasis on the sound management of their employees as we will see in Chapter 5.

4.3 Chapter summary

The highly knowledgeable people in high-tech services firms have a significant impact on the solution delivered either through their industry expertise or through their command of technology.

However, one must note that high-tech services are not always people intensive. Some like consulting or system integration require a significant amount of personnel, while the maintenance or outsourcing services need a balance between people and technology. Finally, network services rely much more on technology than on people.

Furthermore, the various categories of employees within a high-tech services company cannot be considered a homogeneous lot. Indeed, they have significant differences of importance and status according to their relative contribution in the service value, as well as in their relative exposure to the customers.

Personnel is the third key success factor in the high-tech services business for three main reasons. First, employees, such as consultants or maintenance technicians, may have a very significant (positive or negative) impact on customers because they are very often part of the performance experienced by the customer.

Secondly, the experience, the motivation, and the knowledge of employees, such as software engineers or developers, are determinant in the quality of the solution offered.

Thirdly, the costs of people are so important in this industry that they can make or break the profitability of a firm. This is because the core competence of high-tech services firms is less the technology than the people who are using the technology and who are very often in high demand.

Consequently, high-tech services vendors are always looking to have the right number of people with the right competencies at the right time. They constantly monitor their people and performance; they also try to anticipate their needs, both in quality and quantity; and, they seek to keep their people motivated. To do so, they rely on an efficient organization.

In the case of business-to-business high-tech services, the organization of the service providers is significantly different from product oriented firms. There are three main functions (marketing, engineering, and operations) and two major support functions (R&D and human resources).

Regarding the people, there are three different categories of technical populations: the architects and engineering specialists, the operations specialists (such as systems engineers, for instance), and the customer care specialists, who are between the customer and operations.

Most of the successful high-tech services vendors are organized along a supply/demand matrix with a geographical dimension added when those companies are global. They try to constantly manage the balance between being customer focused and being production oriented. They are also putting in place cross-functional teams as a key success factor to building customers' satisfaction (and competitive advantage) through an effective combination of all their various resources.

In the case of business-to-consumer high-tech services, usually vendors go through a three-stage process. First, they start doing some experimenting and testing; the objective is to be there and explore the new opportunities the Web—or other technological innovations—can provide. If the experimentation goes well, the next step is to develop the on-line service activity as a business that delivers an added value to consumers. To do so, the priorities are to build business as well as technical expertise. The ultimate step is to make it an ongoing business and to establish its plain dimension within the organization beside traditional information systems.

Such evolution does not come easily and must be led and backed by top management.

References

[1] Bitner, M. J., B. H. Booms, and M. S. Tetreault, "The Service Encounter: Diagnosing Favorable and Unfavorable Incidents," *Journal of Marketing*, Vol. 54, January 1990, pp. 71–84.

[2] Soat, D. M., *Managing Engineers and Technical Employees*, Norwood, MA: Artech House, 1996.

[3] Kierkowski, A., et al., " Marketing to the Digital Consumer," *The McKinsey Quaterly*, No. 3, 1996, pp. 5–21.

5

Effectively Managing Human Resources

A GOOD ORGANIZATION is a prerequisite, but not a guarantee for effectively leveraging knowledge-based employees to achieve superior performance in the high-tech services business. Serge Kampf, Cap Gemini executive chairman, stated: "It is not that easy to find the thousands of people the group wants to hire this year—half of them to assure growth, half of them to compensate for departures—in a market starved of qualified personnel. And when we have found them, we then have to integrate, train, and motivate them; and it is not hard to imagine what this entails when you are dealing with such strong-minded and highly coveted groups of people."

Consequently, successful high-tech services companies are managing their people very carefully in order to escape the low morale negative loop (see Figure 5.1).

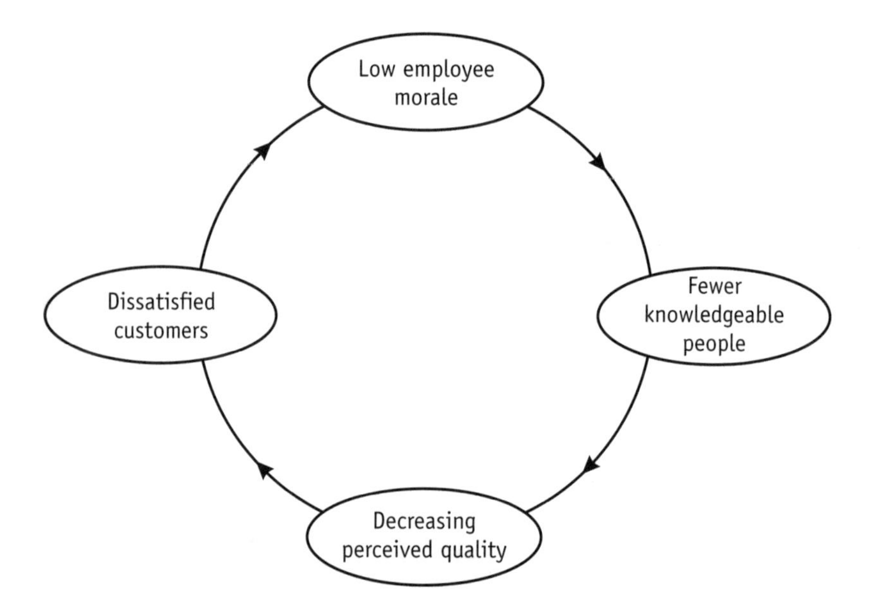

Figure 5.1 The low moral/motivation negative trap.

The importance of employee morale and its impact on the service performance (and consequently on customer satisfaction) has been underlined in various research about people processing services [1]. It is as important in the high-tech services business. Admittedly, technology is also a key driver in the high-tech services value chain, but it needs to be used by people, usually highly skilled people. The best technology available is useless without the right people to exercise it.

It is quite clear that, whatever its origin, low morale in a firm can lead to disastrous consequences for a business. Should it come from salary problems, absence of training, poor motivation or something else, its main consequence is to push employees outside in search of better opportunities. Very often the first employees to leave are the most knowledgeable because they can easily market their expertise and sell it to other vendors or customers.

As a further result, fewer specialists are available and therefore, quality starts declining either in the design of new services, or in the running of current (or existing) services (because of errors, delays, and lack of support). This decline shows immediately in the customers' satisfaction, because they let employees know their feelings (and, hopefully, sometimes the executives of the firms).

If the origin of the problems have not been fixed, such customer complaints add to the low morale of employees, especially the most professional who are aware of the decline of the service they are offering. This can lead to a new row of departures and a new negative cycle up to the point where dissatisfied customers stop using the service and go to competitors with a direct impact on the revenues and the bottom line.

It is usually at this time that top management tackles the original problem, but it is often too late. Customers, as well as the best experts are lost, and it will take time and money to gain their confidence back, much more than if the original problem had been corrected in time.

It is to escape those problems and to create a sustainable competitive advantage that successful high-tech services firms are using the full array of human resources management practices. These include a vision of the role of the employees, as well as their recruitment, training, evolution, management, communication, evaluation, and remuneration [1].

These policies relate to the three main categories of people working in a high-tech services firm, namely: the expert engineers (in technology or in the customers' industries), the technicians (in charge of coding software, maintenance, or running operations and installing equipment on the customer's premise, for instance), and the marketers (mostly account managers). Though these three populations have different backgrounds, experiences, and expectations, the human resources are the same; only the implementation will have to be adapted according to the profile of the people concerned.

5.1 People care as a core value

Successful high-tech services firms have all integrated the "people care" philosophy in the core value of their business as reflected in their mission statements. It does not come as the first element of the mission, but it is always included so that it communicates a very strong message to all employees whatever their position, that they are key contributors in delivering value to the customers. Such an important recognition of their role implies responsibility, but also duty on employees' behalf.

So for instance, EDS asserts: "The second element of our mission is to build an institution able to attract, develop, motivate and retain talented people." The executive vice president of EDS France stated: "We wish for our people to find in their work the opportunities to develop their talents as being the source of pride and accomplishment. We also aspire to be a reference model in protecting our people. In our industry, we are proud to have put in place the first European Supervisory Board, to invest 8% of our total salaries in education, and to have one of the best health insurance systems."

Andersen Consulting's mission statement is "To be one global firm committed to quality by having the best people with knowledge capital, partnering with the best clients to deliver value."

While the CEO of Sema Group concludes its introduction of the 1998 annual report by these significant words: "Finally, Sema Group's strategy depends to a large extent on retaining and attracting high-quality employees. In the past, our challenging projects, and unique culture and image were enough to attract people naturally to join us. However, as the skills marketplace becomes increasingly more competitive, if we are to remain a first choice employer we will have to sell these advantages more forcefully and find additional ways to appeal to and motivate the best professionals in the industry. Sema Group is, after all, the sum of our employees' experience and qualities and it is those professionals who have always assured our success."

A corporate culture is a subtle mix of people's attitudes, corporate history, and past experiences but in many ways it is modeled by top management. This is why successful high-tech services vendors have their corporate executives underline the importance of the people in their business.

This is the prerequisite to ensure that people will be motivated and committed to superior quality in delivering the service value to the customer. It is because employees know that they are cared for that they can care for their customer and their job.

Obviously, the executives must then talk their talk and make sure that the various HR policies are in line with creating this passion for service [2] within their firm.

5.2 The effective recruitment of the right profile

Having the best people starts with recruiting a strong complement of seasoned professionals along with the top graduates from the world's leading universities and colleges. This means that all winning high-tech services companies are investing a significant amount of time and money to recruit.

Money goes into advertising recruitment campaigns in specialized newspapers, as well as to on-campus meetings where the company is introduced to the students and would-be professionals. Very often a specialized section within the company's HR department runs such communication activities.

Time equals the amount of hours spent by managers in interviewing the candidates. This directly concerns the operational managers who will run the majority of the interviews of the applicants for the positions they have open. The HR staff usually conducts only the first round of interviews to make a rough pre-selection and save time for the operational manager.

On average, one position for a consultant may generate 10 qualified applicants (meaning they have passed the HR test) who will be interviewed by an average of four persons (up to six or seven when hiring a strategic consultant) for at least two hours and who will then share their opinions before making a decision. This means that only one new hire takes an average of 100 hours of management time, not including the HR department's time.

Because there is such a lack of available skills, more and more high-tech services companies are using video conferencing to interview candidates, while saving time and money on traveling. Video conferencing allows companies to reach candidates all over the world, a significant advantage for global service vendors. It is an interesting way to prequalify would-be applicants. However, video conferencing interviews cannot replace face-to-face interviews with managers, particularly when it comes to evaluating the potential and various skills of a candidate.

The required profile for any applicant in a high-tech services company first depends on his or her chief competence, should he or she be in charge of operations, engineering, or marketing. However, when interviewing an applicant, leading firms are going beyond the fundamental qualification and looking for the imperative talents needed to achieve success in the high-tech services business, namely, selling skills, business sense, project management capacity, and technical expertise [3].

As shown in Figure 5.2, these skills are shared in various degrees among the employees according to their importance for each category of people. However, since high-tech services for business customers are performed through teamwork, it is critical for each member to possess these capacities in order to deliver value to the customers.

5.2.1 Selling capacity

When dealing with business customers, the selling capacity is the key opener for achievement. In the pre-sale phase (see Figure 3.4) the high-tech services team must be able to identify, understand, and qualify various customer issues; it must then commit the customer to an action plan and a timeframe for developing a solution. It must then sell the solution by showing how it will fulfill customers' needs and then

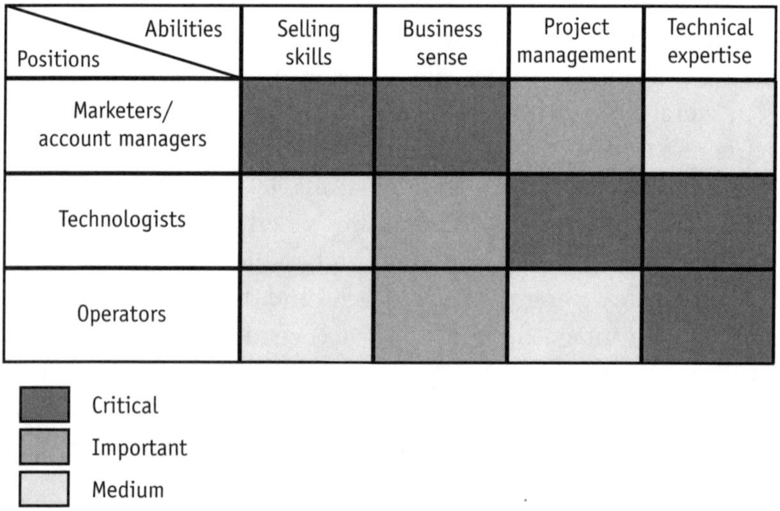

Positions \ Abilities	Selling skills	Business sense	Project management	Technical expertise
Marketers/ account managers	Critical	Critical	Important	Medium
Technologists	Medium	Important	Critical	Critical
Operators	Medium	Important	Medium	Critical

Legend: Critical; Important; Medium

Figure 5.2 The four critical capabilities for highly leveraged professionals.

negotiate the various terms and conditions of the contract, which can lead to a lot of bargaining.

When the delivery and billing is taking place, selling capacities are still required to reassure the customer and to check that the operations are in line with customers' expectations. Accordingly, selling skills are mandatory for would-be account managers and marketers.

Selling skills are also important for the engineering people who are meeting the customer in the pre-sale phase in order to design the solution, because they create a smooth dialog with the customer technical teams. High-tech services are difficult to sell because of their very nature. The customer when buying services is not only strongly affected by the vendor's reputation, but also by the personal chemistry of the account team (including engineering and sometimes operations people) with its own team.

5.2.2 Business sense

Whether or not all high-tech services people need salesmanship ability, they must all have a strong business sense. Indeed, they must be able to understand their customer's important business issues because high-tech services usually have a significant impact on those issues. All employees, should they be in marketing and sales, support, engineering, or operations, must be able to think and talk the language of business, and act, if necessary, to put the customer's business performance first (and their own company second, as we will see in Section 5.5.1).

Providing high-tech services is about delivering top performance through multifunctional teams. The best employees cannot only be experts; they must also have a strong customer focus and a dedication to achieving service excellence by all means. EDS is one company in particular which does this right. One of their core values is that its people must be guided by "an intense desire to do what is best for the client." As we will see later on, such a strong business sense is the cement of international teams who are spread all over the world (see Section 5.5.1).

5.2.3 Project management capacity

High-tech services are complex and demand strong project management skills. Of course, because the technical building of a solution requires the use of different, sophisticated, and intricate technologies, project management ability is a must for anyone wanting to work in the engineering field. Priorities have to be set, activities have to be mapped out, jobs have to be assigned, deadlines have to be made and then progress has to have been watched and controlled, as far as time, quality, and budget objectives are concerned.

Nevertheless, such techniques are also important for account managers in the selling of a solution which may involve a lot of different tasks to be performed by various people at disparate times. In operations, project management is also a consummate way to eliminate hierarchical barriers and concentrate all energies on the completion of process goals.

5.2.4 Technical expertise

Technical expertise is a determining ability of each professional within a high-tech services firm. It is related to the main activity of the employee, whether in sales and marketing, engineering, development, or other support activities. In high-tech services, technical expertise does not mean only having extensive knowledge of the basic skills for a given job. It implies mastery of the latest state-of-the-art tools and techniques available not only in computer, software, or networking technologies, but also in marketing, strategy, quality management, or financial analysis. It also implies a sound and deep understanding of a given industry in order to bring the best solution to the customer.

To acquire such knowledge, high-tech services companies tend to specialize their people by industry, either when they join the company or in a progressive manner as done at Andersen Consulting where people are exposed to different categories of customers when they are in junior positions, and then are aligned gradually in a given industry.

In order to maintain and develop the knowledge capital of their employees, many successful high-tech services vendors are using sophisticated knowledge booster tools. For instance, companies such as IBM Global Services, Andersen Consulting, Hewlett-Packard, Compaq, and Cap Gemini are globally connecting all their service teams to an electronic knowledge sharing Intranet system. Such a system allows their people to have access to the skills and experiences of all the firm's professionals, along with information-rich databases and other resources. They are also providing their people with a rich variety of training.

5.3 Cogent training

All leading high-tech services vendors are spending a significant amount of money and time on education for their people. For instance, in 1996, for its 45,000 professionals worldwide, Andersen Consulting invested an average of 120 training hours per person for a total of 5.4 million training hours at a cost of $332 million. This represented 6.5% of revenues and about half of the profit!

Such a cost is important and it pays back. Training allows each professional to hone technical and behavioral skills and to develop new ones. It does not concern only newcomers, but everyone in the company. Actually, the most senior people are sometimes the most in need of training because competence has to be continually updated and developed in the high-tech services business.

Training is also a key tool to managing change within a company by re-orienting its competent resources internally. An executive of Compaq said: "We need to develop competence perpetually; this means a large effort in training. Besides, there are intellectual limits to this business and we need to adapt our people." Considering the shortage of skilled people in the market, companies are trying not to waste rare resources. Not only does high impact training show a firm's commitment to its employees, but since their main asset is their knowledge, effective training enhances their market value, hence their satisfaction.

But training is far more than formal sessions in the high-tech services business; it is a continuous learning process for all professionals. Accordingly, on-the-job training is also widely organized and heavily used in leading firms. It relies on peer and manager reviews. All the major firms are holding account reviews, process reviews, and project reviews, which allow team members to share the elements and the lessons learned from these various experiences. In order to build an ever richer base of knowledge and intelligence, IBM Global Services has a special customer review on major projects performed by an external consultant which is then presented to and commented on by the various team members.

For newcomers and junior employees, another effective on-the-job training is mentoring or coaching by senior employees. Some companies have put in place a so-called "buddy system" where a senior staff member will take care of a younger or less experienced one. In other firms, a coach may follow between 5 and 10 people, outside of their direct reporting line. The coach is a point of contact for discussing all sorts of problems; the coach also acts as a role model and provides junior employees with tips and advice on technical or behavioral matters.

As we will see in Section 5.7, mentoring also plays a key role in the evaluation process. Mentors are motivated to perform well because it is part of their job description and their own evaluation review will include a judgment of their coaching ability. In other words, they will develop themselves by developing others.

5.4 Professional development

Gone is the time where the motto in high-tech services firms was "Up or Out," meaning that any professional delivering an "average" or "above average" performance in a given position should go after a while because he or she could not find their way in the organization. While this had the advantage of putting pressure on people, it also led to numerous cases of burnout, as well as to high turnover.

That was acceptable when business was limited and resources were overflowing. Since the middle of the 1990s this is no longer the case. Facing a booming business but a clear lack of enough skilled people, companies are anxious to keep their best employees.

"Up or Out" leaves the room to "Grow or Go," meaning that a professional must constantly enlarge his or her capacities (selling skills, business sense, project management, technical expertise in sales and marketing, or engineering or operations) but does not need to climb the corporate ladder to achieve status, recognition, and reward. One executive of Andersen Consulting mentioned: "Promotions are made at various speeds according to the performance and expectations of each of our staff."

Many professionals are not keen on switching to a management position that requires specific skills and always signifies the loss of hands-on technical expertise. Today, companies prefer to keep those professionals and offer them various career paths. For instance, IBM Global Services puts a strong emphasis on functional mobility (i.e., moving horizontally from one function to another) as well as international opportunities.

5.5 People management

Before training and development, the management of people is a key driver of the behavior of employees because it relates to their day-to-day activity. Most of the jobs are stressful in the high-tech services business because of the complexity of the projects, the pressure of the competition, the intricacy of technology, and the demand of large corporate customers who often have important business risks at stake in a high-tech service. To perform well under pressure, professionals need to be backed by a sound management system. According to a famous study in the beginning of the 1980s [4], performance troubles are present in more than 75% of the cases from management problems and not from the individual difficulties of the performer. Bad management is often the ignition point of the low morale negative loop as seen in Figure 5.1.

The solutions are twofold: full empowerment and clear performance specifications.

5.5.1 Full empowerment

In many flourishing service firms, empowerment of the employees, most notably the contact employees, is seen as an essential element to providing customer satisfaction. Probably, one of the best rationales for empowerment is the one given by Jan Carlson, former CEO of Scandinavian Airlines: "To free someone from vigorous control by instructions, policies, and orders, and to give that person freedom to take responsibility for his or her ideas, decisions, and actions is to release the hidden resources that would otherwise remain inaccessible to both the individual and the organization" [5].

Empowered employees are certainly more customer centered and when facing customers, are more sensitive to their needs in knowing they can find and design a specific service or customize an on-the-shelf solution. Since they have latitude to behave, they can react more quickly and positively to a service failure, knowing that the recovery of a dissatisfied customer can reinforce satisfaction and loyalty.

In the high-tech services business, empowerment is not only a solution for recovering from a service failure at the operation level (as considered in more traditional service firms), but it is also an efficient way of stimulating creativity for devising and designing new innovative solutions in marketing and engineering departments.

One executive of Atos said: "In my branch, I have a lot of engineers. We favor autonomy, small teams in charge of a customer from A to Z, and the appeal of being at the forefront of technology, like with the Internet, for instance." Similarly, according to a director of Cap Gemini: "We delegate a lot of responsibilities to the sales and technical teams in charge of the relationship with a customer. They manage their business, their own people, and their P&L by job or customer."

Empowered firms have a management system that is very different from the production-line approach, such as at fast food companies where employees are taught how to perform every step of the service and to act by the book [6]. Such an approach secures quality in low cost, high volume, people-intensive service operations [7] but does not work in high-tech services, which are low volume, high cost, high-tech services.

Even in the case of consumer on-line services, such a practice does not work because the mass of operations are performed through electronic (and not human) interfaces, while the challenge for on-line services vendors is clearly to stimulate the creativity of their engineering and marketing people to imagine attractive and efficient on-line service applications.

Empowerment control systems focus on the performance measured by actual outcome and not on the behavior of people subjectively evaluated by managers. What better symbol of empowerment than the disappearance of the formal dress code for the IBM salespeople? The dress code worked well to sell large computers and to convey a standard image of seriousness, power, and control. Today's customers, however, expect their high-tech services suppliers to be creative, efficient, and committed to delivering the best performance, whatever the look of the vendor's people.

Empowerment means that the manager acts much more like a tutor than like a controller. Being in charge of very skilled people—often with a strong potential to grow—the job of middle management is to identify star performers, help other professionals to increase their competence or help them to leave, and to be a facilitator for getting the necessary resources available to the team.

Regarding top managers, their job is less to manage their manager—along the command and control chain, such as in the production line—than to give clear strategic direction for the company, as well as to provide the right setting for empowerment to work. This goes through a culture of trust and respect for the people.

Trust is a two way process. When senior management trusts the staff, it is reciprocated. Trust reinforces the loyalty of the employees and decreases the turnover.

Furthermore, a culture based on trust makes for bigger and faster expansion. For instance, Andersen Consulting could not have doubled its revenues and expanded the number of its people by more than 66% in the past 5 years if it had controlled everything (see Figure 5.3).

Trust is also the best way to manage multifunctional teams including account managers, solutions architects, project managers, software experts, quality engineers, etc.

Trust is the only way to efficiently manage the virtual teams that work together for a limited amount of time on a global project at various geographic locations. For instance, one executive at EDS gives the example of how EDS managed to provide one Japanese customer with

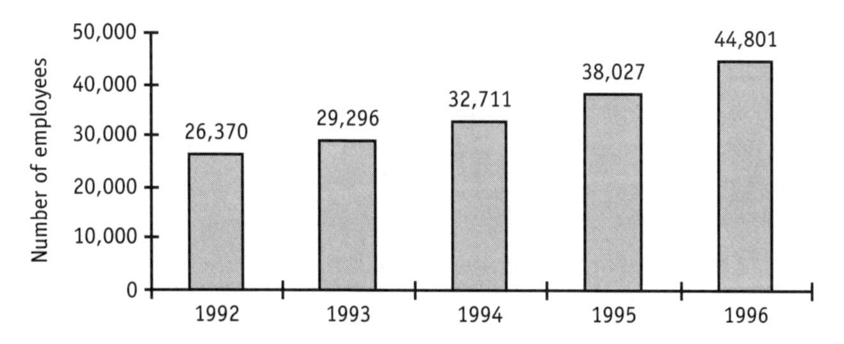

Figure 5.3 Andersen Consulting: Total personnel worldwide.

a communication system for its Malaysian subsidiary in order to introduce common business processes between the firm with its stores, shopping centers, and distribution points across Malaysia. "The account manager was a Singapore citizen based in Kuala Lumpur; on the technical side he was teaming with a Hawaiian-born engineer who had gained previous experience working for one of our Swedish retail customers. A Malaysian engineer supplied the information technology expertise while a cost analyst, American born and based, oversaw costing activities. Legal negotiations were driven by one of our Australian lawyers. This entire virtual team was managed from Japan by a Japanese manager." Such international and geographically dispersed teams are very much the rule in all major high-tech services vendors with a global scope.

What makes such teams successful is not control, but an intense desire to do what is best for their clients, as well as a willingness to deliver quality service through international friendship. It is up to managers to fuel those feelings by backing such teams with the necessary resources in order to allow their people to perform [8].

The down side of empowerment is its costs because the firm must check that the right people are empowered through careful recruitment and training. However, this is not a real problem for high-tech services firms due to the kind of profile they need to hire because of the nature of the business. Once they have managed to attract people with high potential or top performers in their field, they must give them free rein so that they can express their full capacity.

However, empowerment does not come by chance. It needs the intensity and the sincerity of top management to organize systems that champion empowerment through the whole structure of the firm, so that it becomes a reality and not a mantra. It also needs to be reinforced by training that is designed to increase the staff's capacity to use empowerment for maximizing customer satisfaction and not its own personal goals.

5.5.2 Clear performance specifications

In return for enabling their people, successful high-tech services vendors are also making their performance specifications very clear, based on quality, time, and budgets. They are asking each of their professionals to optimize their resources in order to deliver the most time efficient and least expensive way to meet the performance specifications defined by customer's expectations.

Those firms couple extensive documentation of previous experience with a deep understanding of existing customers so that they can set (and negotiate) sensible benchmarks for performance standards. For instance, EDS has created a tool (Knowledge Office™) running on EDS's internal information network, which arranges, displays, and updates information that people use more often on a worldwide basis. The program gives quick access to the nature of the information, its source, and its context of use. Such a knowledge tool is a great productivity booster because it saves users a lot of time, as well as being a good indicator for any benchmark reference.

High-tech services vendors also rely on an intimate knowledge of their own cost structure and productivity measures [8]. They are then taken into consideration in setting specifications to be sure that the solution will also meet the profit expectations of top management.

Ultimately, performance specifications are understandable, measurable, and achievable. They are usually made at the customer team level and then split by function (sales, engineering, operations), by project, and by person whenever possible.

5.6 Internal communication and feedback

To help the staff know if it is in line with the performance expectations, high-tech services companies provide constant feedback with status of quality, time, and budget. Since the most relevant and fundamental feedback comes from the customers, formal meetings are scheduled on a regular basis to review the status of current projects, as well as to review the functioning of the entire customer-vendor relationship and organization.

As we have seen in Chapter 2, those meetings are necessary for ensuring the quality of the services delivered. But they are also a very effective management tool to help the vendor's professionals to monitor their own fulfillment against the performance specifications and the client's expectations.

Managers show tracking charts about customer satisfaction, as well as productivity measures (projected hours vs. actual hours, projected budget vs. actual budget, percentage of project actually achieved vs. planned, and so on). A manager at Cap Gemini stated, "Tracking charts displayed prominently are powerful transmitters which orient behaviors and attitudes."

Periodic face-to-face status meetings are an effective method for refreshing and motivating people as well as teams. They are also the place for sharing information, discovering potential problems, and developing contingency plans just in case.

Internal and external audits are also very effective tools to provide feedback to employees because the audits evaluate the existence, the effectiveness, and the efficacy of the processes and procedures implemented. Very often such audits introduce fresh ideas, stretch the thinking, challenge the existing practices and are also a good way to develop less experienced staff on the job.

More generally, successful high-tech services vendors have put in place an effective internal communication system to help management to shape the collective culture and commitment of each staff member to achieve superior performance and deliver premium value to their clients.

All major firms rely on the traditional in-house communication tools such as information newsletters, posters, and videos, but their preferred device is the internal information network. Not only does such a network, usually a global Intranet type, give access in real time to a flow of information; but it is also a vital tool to keep the various members of the virtual teams in touch as we mentioned above. Without such a network, staff members spread over various geographic locations would not be able to work together.

5.7 Evaluation and remuneration

Today, all the leading high-tech services vendors reward their people by offering them salaries that are clearly above the average on the market. In addition to that, they are tying their compensation system to customer satisfaction and the ability to deliver value efficiently. They can manage to do so because clear performance specifications pave the way to effective evaluation and remuneration.

Because of the nature of the business and the importance of multifunctional teams, whenever possible, firms are linking bonus compensation to team performance, creating a strong incentive for each individual not to let the team down.

However, money and incentives—such as trips or gifts—are not the only motivation for people in the high-tech services business. Promotions are also a traditional and effective way to acknowledge and reward outstanding performance. The executive vice president of Syseca stated: "We identify and monitor our high potential people and have a yearly people review." Nevertheless, the number of top management positions is by definition always limited and nowadays, companies tend to prefer a "Grow or Go" attitude instead of the old fashioned "Up or Out" to strengthen their employees' loyalty and decrease turnover.

Such a move actually works well because the personal pride of job achievement is a very strong motivation for employees. This is especially true when being successful is a complex job involving state-of-the-art technology for a large customer creating a multimillion dollar impact on the customer's business. Accordingly, star performers are publicly honored either at annual dinners or in the in-house press and networks. The firms also provide achievers with the opportunity (most notably in engineering or operations) to write and publish articles, to give presentations at various conferences, and to attend professional meetings to meet other peers in the same field.

Similarly, offering training is a strong incentive to allow people to stay at the top of their expertise and to continually expand their knowledge capital. In the high-tech services business, knowledge is what gives credibility, consideration from peers and colleagues, as well as market value. Smart companies know that and try their best to nurture the learning of their people.

5.8 Chapter summary

Successful high-tech services companies are managing their people very carefully to achieve superior performance. Facing a booming business but lacking skilled people, companies are anxious to keep their best employees. They manage to keep employee moral and motivation high through the full array of human resources management practices. These include a vision of the employees' role, as well as their recruitment, training, evolution, management, communication, evaluation, and remuneration.

They have all integrated the "people care" philosophy in the core value of their business as reflected in their mission statements. It communicates a very strong message to all employees that they are key contributors in delivering value to customers. Such important recognition of their role implies responsibility, but also duty on employees' behalf.

They are also investing a significant amount of time and money to recruit seasoned professionals along with the top graduates from the world's leading universities and colleges. The required profile for any applicant first depends on his or her chief competence, but companies even look for broader potential capacities, namely selling skills, business sense, project management, and technical expertise.

Moreover, companies are spending a significant amount of money and time on education for their people. Training allows each professional to hone technical and behavioral skills and develop new ones; it is also a key tool to manage change within a company by re-orienting its competent resources internally. On-the-job training is also widely organized and heavily used in leading firms; it relies on peer and manager reviews.

Regarding professional development, the principle of "Up or Out" leaves the room to "Grow or Go," meaning that professionals must constantly enlarge their capacities (selling skills, business sense, project management, technical expertise, sales and marketing, or engineering, or operations) but do not need to climb the corporate ladder to achieve status, recognition, and reward.

Most of the jobs are stressful in the high-tech services business because of the complexity of the projects, the pressure of the competition, the intricacy of technology, and the demand of large corporate customers who often have important business risks at stake in a high-tech service. To perform well under pressure, professionals need to be backed by a sound management system, which rests on full empowerment and clear performance specifications.

Empowerment is not only a solution for recovery from a service failure at the operation level (as considered in more traditional service firms), but it is also an efficient way of stimulating creativity for devising and designing new innovative solutions in marketing and engineering departments. Empowerment control systems focus on the performance measured by actual outcome and not on the behavior of people evaluated subjectively by the managers. Empowerment means that the manager acts more as a tutor than as a controller. This goes through a culture of trust and respect for the people.

As the result of this empowerment, successful high-tech services companies are presenting their performance specifications, based on quality, time, and budget and they provide constant feedback through status reviews and other internal communication tools, notably an internal real-time information network.

Finally, successful service companies reward their people by offering them high salaries and attractive compensation tied to customer satisfaction and delivered value. Promotions are still a traditional and effective way to acknowledge and reward outstanding performance but they are not as important as in more traditional service businesses. Recognition and training are also strong incentives in an industry where knowledge is the ultimate value.

References

[1] Soat, D. M., *Managing Engineers and Technical Employees: How to Attract, Motivate, and Retain Excellent People*, Norwood, MA: Artech House, 1996.

[2] Benjamin Schneider, B., J. K. Wheeler, and J. Cox, "A Passion for Service: Using Content Analysis to Explicate Service Climate Themes," *Journal of Applied Psychology,* Vol. 77, No. 5, 1992, pp. 705–716.

[3] Alexander, J. A., and M. C. Lyons, *The knowledge-based organization: our steps to increasing sales, profits, and market share,* Burr Ridge, IL: Irwin, 1995.

[4] Deming, W. E., *Out of the Crisis,* Cambridge, MA: M.I.T Cambridge Center for Advanced Engineering Study, 1986.

[5] Carlson, J., *Moment of Truth*, New York: Balligen, 1987.

[6] Zemke, R., and D. Schaaf, *The Service Edge: 101 Companies that Profit from Customer Care,* New York: New American Library, 1989, pp. 68–69.

[7] Bowen, D. E., and E. E. Lawler III, "The Empowerment of Service Workers: What, Why, How, and When," *Sloan Management Review*, Spring 1992, pp. 31–39.

[8] Kuhlken, L. E., *Expanding professional services: a manager's guide to a diversified business,* Homewood, IL: Business One, Irwin, 1993.

6

The Marketing Strategy for High-Tech Services

MARKETING FOCUSES on making a product or a service available at the right place, at the right time, and at a price that is acceptable to customers [1]. A more detailed definition is provided by the American Marketing Association (AMA): "Marketing is the process of planning and executing the conception, pricing, promotion, and distribution of ideas, goods, and services to create exchanges that satisfy individual and organizational goals" [2].

One must notice that this definition must be adapted to the case of the high-tech services because of their singularities. Indeed the "marketing mix" for high-tech services is slightly different than for product or traditional services.

Most notably the distribution strategy is very limited. In the case of business to business marketing, all the high-technology services vendors are directly selling their services "because of the limited number of customers and because (service) quality cannot be delegated to a third party," said one vice president of Syseca. In the case of consumer on-line services, the only distribution channel to sell and deliver is the Web and there are no third parties.

Regarding the conception of new services, it is interesting to notice that high-tech services allow the development of solutions in line with the principles of relationship marketing. They offer flexibility, which makes them more easily customizable than traditional services.

Moreover, the communication strategy, notably concerning the brand image, is very important not only because what is sold is intangible but also because the service can significantly impact the activity of a customer. When asked to describe the communication strategy of IBM Global Services, one executive answered: "Our communication relies on:

- The promotion of the brand IBM Global Services which identifies our range of services all over the world;
- The highlights of facts which are both actual and checkable so that our customers can verify our ability to perform and deliver. Service cannot be sold in the same way as a product. It is sold through the experience of the provider because it is intangible. Accordingly, one must say what it makes and makes what it says. It is a question of credibility;
- A strong segmentation of our target customers in line with our strategic goals."

Such statements clearly stress the key element of a sound communication strategy, namely the importance of the brand image, the effective communication of the service value to customers and the precise definition of the target audience.

Finally, the pricing strategy for high-tech services is not very different from traditional services apart from the important content of innovative technology within those services which need to be taken into account.

Let us now examine in detail the various adaptations of each variable of the marketing mix of high-tech services.

6.1 Achieving relationship marketing and conceiving new high-tech services

One of the buzzwords of the 1990s, *relationship marketing*, emphasizes the importance for firms to organize a continuous connection with existing customers before trying to attract new clients [3]. It emphasizes "after-marketing" [4], that is, all the actions after the initial sale, as a key component of marketing. Relationship marketing encompasses all the operations aiming at boosting customer loyalty (see Chapter 2) and recovering customers in case of service failure (see Chapter 3).

But relationship marketing goes beyond customer retention. Its objective is to differentiate the service offering of a firm from those of competitors on dimensions that are meaningful to customers and difficult for competitors to duplicate [5]. The ultimate goal of relationship marketing is the one-to-one marketing where each segment of the market is only one customer, meaning that the firm is able to deliver a specific solution to each of its customers [6].

This is clearly the case for professional high-tech services because of the limited number of clients. Leading high-tech services firms are running their best customers as part time employees for generating new ideas. They know that the best customers are frequent service users who know a lot about the services and are often the source of innovation. Actually, all the successful high-tech services vendors are already organized by industry with a dedicated account management structure.

But one-to-one marketing also works for on-line services to consumers. As we have seen in Chapter 2, it is easy to interact directly with users on the Web and to tailor the offer to the profile of each customer using "cookies" or other techniques to identify up front the profile of each customer [8].

Relationship marketing has dramatic consequences relating to the development of new services. One executive of IBM Global Services stated, "Building service offers with a stable content is a key element of our marketing strategy. The domains for potential offers are identified through our current business analysis and market studies. We want to market offers that are consistent and available on a worldwide basis; this implies major investments to launch them. Consequently, the decision is made at the corporate level, very often after a local pilot test which is used then to spread the knowledge."

A marketing manager at Hewlett-Packard confirmed, "New services are made according to the information coming from the various countries to our world center for service creation. We rely on surveys of our customers and sales representatives, as well as test customers and pilot customers. Once a need has been identified, we look for a solution both homogeneous and serviceable in every country while including the opportunity for a local adaptation."

Such a way of creating new solutions balances the necessity of fulfilling each customer's needs with the imperative of relative standardization to achieve quality and productivity (see Chapter 3).

Indeed, high-tech services fit very well with the model of mass-customization, this golden fleece of marketing which aims at satisfying each customer while enjoying the economies of scale of mass production. The very nature of high-tech services which are relying on a flexible technology allows the addition of any specific request from a client to the core system of the service. Such a customization is much more difficult to achieve with high-tech products because of the limitation induced by the hardware. It is easier, faster, and cheaper to write some extra lines of software code than to design, produce and distribute a specific device for say desktop PCs which are already among the more mass customized product.

Actually, high-tech products and services are looking for the same goal, mass customization, along two different paths. Because of the manufacturing constraints, high-technology product vendors offer standardized products they try to customize by offering various attractive options to customers. On the contrary, high-technology service vendors start usually from a very customized solution, designed for one customer, and try to duplicate this service as much as possible in order to achieve productivity and improved quality through economies of knowledge (see Figure 6.1).

But high-tech services, like high-tech products, include a significant amount of technology, which is not without consequences on the creation of new service offerings. Whether some new services are clearly market driven, others are pushed by the technology.

One executive of Andersen Consulting said: "Sometimes customers call us and ask for a specific solution. But we also define market offering

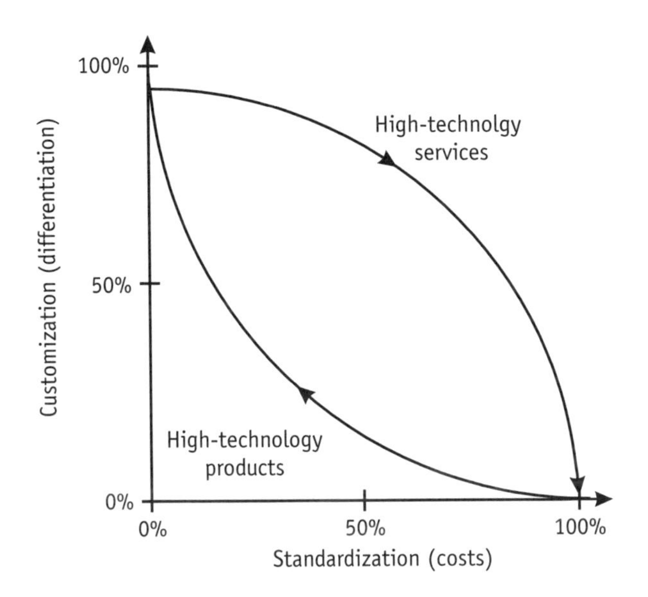

Figure 6.1 The conception and implementation of new high-tech services.

by conceptualizing our knowledge on hot topics regarding organization, process, or technology." Similarly, a vice president at Atos, one of the leading European providers of high-technology professional services, proclaimed: "In my division, there is a real offering driven marketing based on the latest technologies available whose possible use we try to identify on potential clients. This job and the investigation of test customer is managed by a small marketing team and by an R&D team of 25 people, with one project manager in charge of a leading technology."

Such offering driven marketing relies generally on the " beta testing" of service prototypes by selected test customers. If the test is positive, the service is then marketed to a wider segment of similar clients with similar needs. Clearly the approach fits with the mass customization model which starts from a specific offer and intends to generalize it.

Sometimes, the offering comes from an in-house experience since all the high-tech services vendors are using the very services they are proposing to their customers. For instance, one executive at Hewlett-Packard mentioned: "If we have developed our own services offering, naturally it is because our customers have requested it and because this competence helps us not to be considered as a trifling computer box maker. But it is also because we are our own test laboratory with more than 100,000 workstations in the world connected by a gigantic Global Intranet/Internet network. For instance, once we have developed and fine tuned solutions internally such as help-desk or on-line software distribution and updating, or remote network monitoring, we can offer them to our clients."

Whatever the origins of a new service—in house or from customers—all leading high-tech services firms have roughly the same development process (see Figure 6.2). It usually starts with the generation of ideas coming from customers which then are stored in a service creation database. Typically such a database may have between 500 and 1,000 ideas which are kept alive. Some stay as ideas because

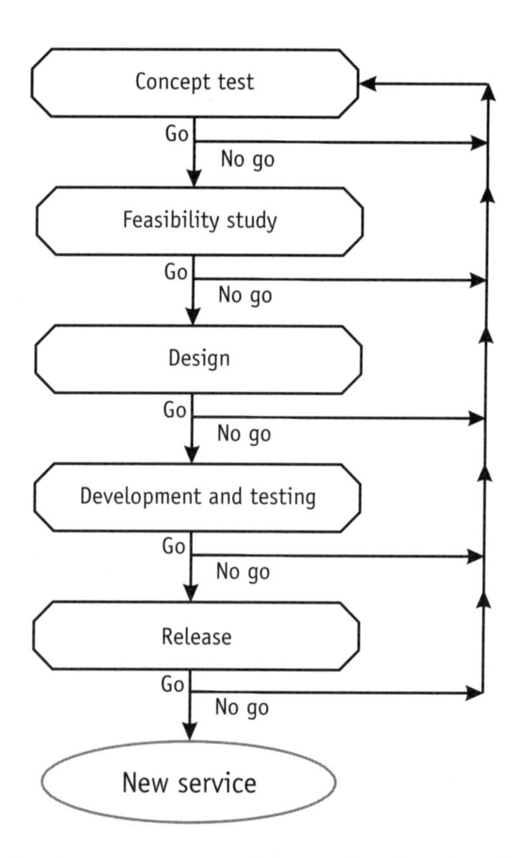

Figure 6.2 The development process of new technology-based services.

the environment is not yet ready (from a technological, political, or economical standpoint); however they are scrupulously kept because business conditions may evolve quickly and allow them to be translated into a service concept.

The service concept describes the key benefits of the proposed service for the customers and is presented face-to-face to would-be customers either individually or in focus groups or even virtually through the Web. The concept test assesses the reactions of the customers that can:

1. Not understand the service;
2. Understand it but are not interested;
3. Be mildly interested;
4. Be ready to experiment with it.

If customers seem interested in the concept, then the value of the service is evaluated in comparison to the existing services and those from the competition. At that stage, the matching of such a service with the corporate strategy and the service portfolio is assessed. Its business and strategic value is also pondered, notably by estimating the potential demand and revenues.

When the concept seems worthwhile, the following phase is to assess its feasibility, mostly from a technical point of view but the legal aspects are also taken into consideration, especially for Web services as well as the patent issues. This phase is performed by a multifunctional team under one project manager specifically assigned to this new service in progress. At that stage, the group goes through the knowledge databases to try to identify whether a part of the solution has already been developed or could be reused for the new service. Everything that needs to be developed is estimated at that stage which is concluded by the writing of a technical feasibility assessment report.

After the report is positive, the time arrives for the design phase which concentrates on the technical and operational requirements of the would-be service. Various solutions may be considered, either in-house or externally. A detailed plan is issued, as well as a service blueprint which describes the entire service process in order to identify the crucial parts of the service. The design phase is completed by a marketing plan and a business case with the revenues and costs associated with the future service, so that the directors in charge of development can make a decision to go on or not.

If they give their approval, the project enters the development and testing level. The service is usually developed and tested first internally (Alpha test) before being tested by selected customers (Beta test). Customer feedback is integrated and modifications are made to refine the service. Once the period of testing is achieved, the business case is updated and the decision is made whether or not to release the new service.

When the release is approved with a date for launching, it is time to prepare communication for the announcement of the service, carry out in-house training for the related employees, and implement the logistic operation associated with the service. Then comes the launching of the service and the critical task is to evaluate customer awareness and evaluation of the new service. Having a strong image for a high-tech service clearly facilitates its launching.

6.2 The importance of brand image

A brand is a name, a set of words, a sign, a symbol, a design, or a combination that identifies a vendor's product or services. In the high-tech services business, branding is an absolute necessity for many reasons (see Figure 6.3).

First, adding a brand name to a service changes its essence and adds a psychological value to the service. An executive of Andersen Consulting stated: "Brand image is very important in our business because there is no product. We are selling intangible promises where we translate our expertise, our competence, and our value. A strong image encompasses all those elements."

Secondly, a brand name helps customers to make their decision by saving time in their search for information. According to the executive vice president of Syseca: "Branding is very important because the short-lists are more and more condensed and notoriety is a key success factor to be requested for a proposal or a bid."

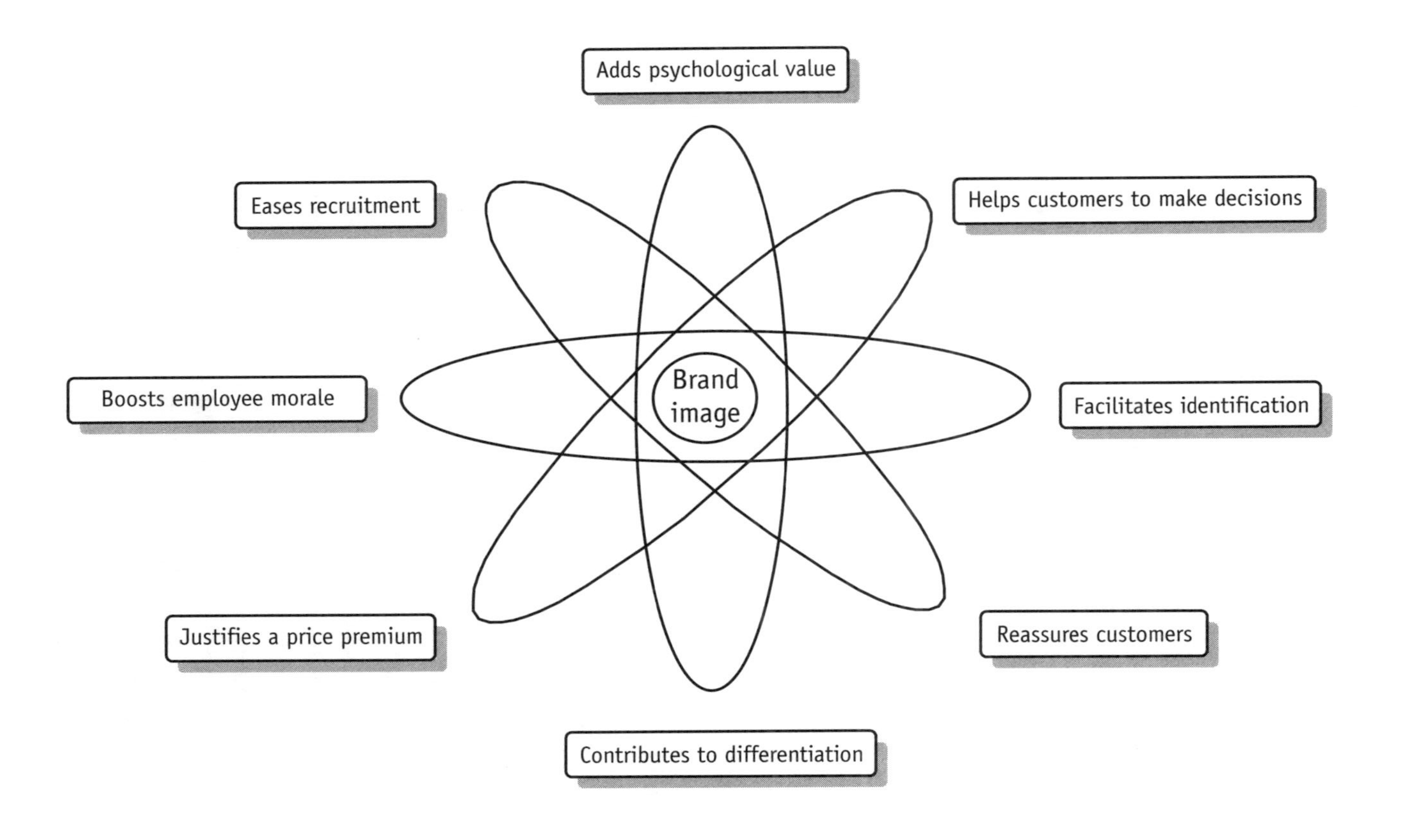

Figure 6.3 The eight benefits of a strong brand image for high-tech services.

Furthermore, a well known and famous brand image reassures customers, an essential criteria when making the decision to use a high-tech service which often is no more than a leap into the unknown.

One executive of Atos said: "A strong brand image is fundamental in our business regarding our target market: large suppliers reinsure large customers. Moreover, nowadays our clients are outsourcing large blocks of their activities to high-tech services firms, which need to be even more credible in order to get the business."

One manager of SG2 added: "We are often working on strategic issues for our customers. A strong image and good references inspire confidence."

Another benefit of branding is to facilitate service identification while attaching a quality image and a personality to the service that builds customer loyalty and justifies a price difference. One executive of the service division of Compaq affirmed: "Brand image stands for an identification with the level of quality and security that we are offering and which is acknowledged by the market."

Identification is of the foremost importance when the company is proposing various products and services to the markets such as the computer companies (IBM, Compaq, Hewlett-Packard, etc.) or the package software companies (SAP, Banian, etc.). According to one executive at Hewlett-Packard: "The brand image is important because it allows customers to discover the offering of services by Hewlett-Packard, which already enjoys an excellent image of quality for its hardware platforms."

Because it encompasses a psychological value as well as an identification of the service, a brand image makes it easier for the customer to differentiate the solution it represents from the solution offered by other competitors. Such a benefit is clearly essential when services can be either commoditized (as in the case for some basic network services such as network management services) or easily imitated like many Internet-based services. For instance Yahoo, the leading search engine, has differentiated from faster (and maybe better) competitors through a cool, likable (and now well known) brand image. In 1998, the Yahoo site was the most attended site on the Web with more than 30 million hits.

Branding is also essential for high-tech services firms because it has a significant impact on employees. First, it contributes to motivating the staff who feel proud to work for a well-known and admired company.

Such a feeling is very often the actual mortar for people working in multi-functional teams, especially if they are spread all over the world. Brand image is a key constituent of the firm's identity.

Furthermore, a strong brand image helps to select new recruits because it enlarges the number of applicants and it attracts the best candidates. This is not the least important benefit of branding at a time when it is difficult to attract new talent in the business, as is the case nowadays.

Conversely, a bad image can have disastrous consequences not only towards the clients but also within the firm. A company with a bad name has difficulty retaining its best people and attracting new ones. A bad image may start or feed the low morale negative loop described at the beginning of Chapter 5.

For these reasons, building a strong brand image creates a real value addition to customers, which translates into brand equity for the vendor.

The value of a brand is measured by the degree of its awareness in customers' minds. Typically, a powerful brand will go through the following stages: From *zero awareness* to *assisted recognition*, when it is mentioned in a list of brands submitted to respondents; to *unaided recall*, meaning that the respondent will associate the brand name directly to a given product or communication message; to *top of mind,* when the brand is mentioned first without any assistance.

A strong branding strategy is based on three key principles: dominance, exclusivity, and singularity. Usually in the thought sequence of a customer, the service is identified first, then the brand comes to mind. Consequently, a dominant brand is the one that comes to customers' minds first, before competitors. A research program developed by the U.S. Strategic Planning Institute (SPI) in the late 1970s indicates that such dominant brands have greater returns than their competitors. On average, the "top of mind" brand has a return on investment (ROI) of 34%, while the second competitor has 21%, and the third 16%.

An exclusive brand is a must, because experience and research show that two brands cannot both occupy one position at the same time. Even worse, any major communication investment by the second brand usually reinforces the leader's position with customers by making the association more salient. AOL has been a big winner in that game.

Finally, a brand cannot occupy two distinct positions at the same time in customers' minds. Such was the problem with Andersen Consulting, which was perceived much more as an accounting firm than a high-tech services provider. To many corporate customers, "Andersen" was more like Arthur Andersen, the famous accounting firm. Similarly, IBM Global Services had to struggle to convey an image of a service provider when many customers thought of IBM chiefly as a high-technology product vendor. Such is also the case for the service divisions of Compaq and Hewlett-Packard.

But a strong brand recognition also means a significant amount of money is invested to promote the brand up front. The human mind does not build up favorable impressions slowly over time (see Figure 6.4). Usually, once a customer's mind is made up, it rarely changes and a perception that exists in the mind is often interpreted as truth. Consequently, a strong branding strategy for a new product or technology requires a "big bang" to establish an initial position in customers' minds; only then can subsequent input strengthen and sustain this first impression.

Most of the successful high-tech services companies have made the decision to promote their corporate brand name, using it as an "umbrella" for the various categories of services they are offering. The latter may have their own brand name, but it is mostly for the sake of

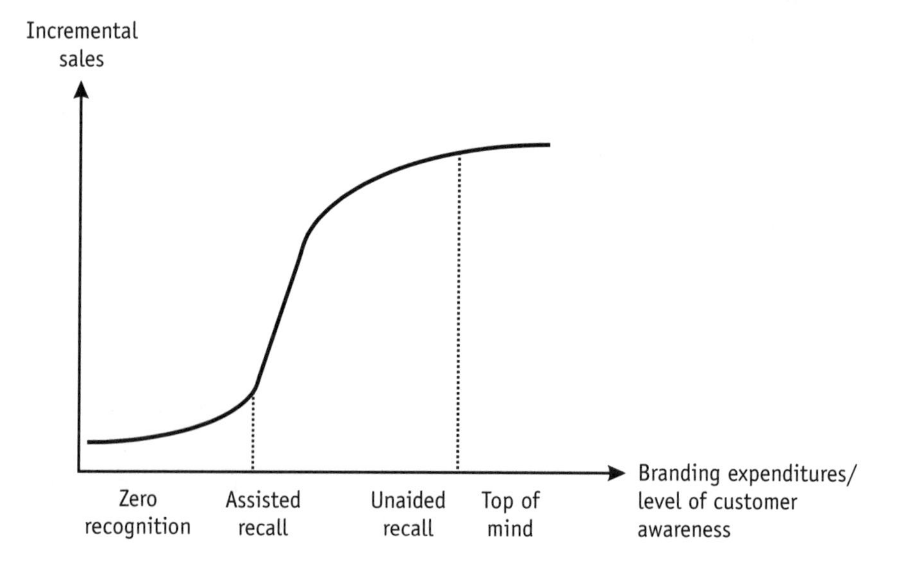

Figure 6.4 The "S" model of custom response to brand awareness.

identification that they are always attached to the company's name. Even in the case where those services are offered by the division of multibusiness firms, such as IBM or Hewlett-Packard, those divisions take great care to publicize their own "corporate" brand names, such as "IBM Global Services" or "Hewlett-Packard Software and Service Division" to identify their offerings under a generic cover.

Such an umbrella strategy is driven by the fact that in the high-tech services business, the brand image has a dramatic impact on the internal organization, which then snowballs in external performance. Consequently, all successful high-tech services firms have managed to build their image and their fame through a strong internal culture.

In the business of services, if a company is not able to deliver value, then advertising a brand name is purely cosmetic. The gap between the projected image and the reality perceived by the staff may create negative reactions from the employees.

Consequently, the brand image must be managed externally but also internally and all the communication activities are reflecting the firm's core values as well as its competencies. Most notably, considering that advertising messages are received both by customers and employees, the latter may respond negatively if an advertising campaign oversells promises they are unable to deliver.

The best way to develop a strong brand identity and to defend its territory, is to start with the firm's identity and to define the brand identity before designing the instruments to show this identity, such as company and brand name, trademark, advertising, sponsoring, direct marketing, etc.(see Figure 6.5).

By focusing on its core values, which are modeling its performances, the company decreases the risk of discrepancy between the communication and the actual delivery performance of its service offering. It also strengthens the culture of the organization, making the firm more effective internally and more credible externally.

Service companies with the best brands in the world [9] share some similar features:

- They are constant in communicating to their clients. They send coherent and consistent messages over years, although they endlessly refresh the way they communicate the brand image.

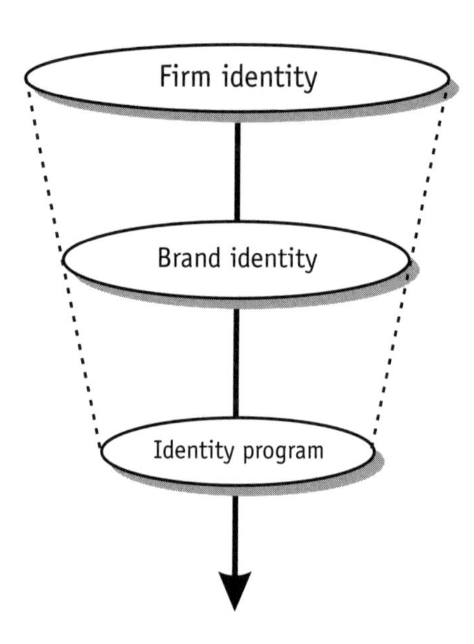

Figure 6.5 The process of building an effective brand image for high-tech services.

- They own an exceptional image, clearly different from their competitors and appealing to their customers.
- They communicate the total value for customers instead of a specific dimension of the service.

Those three characteristics are clearly shared by some of the most successful high-tech services companies.

Ultimately, a strong brand name comes from matching the image set by the firm with the customer experience. This is the reason why high-tech services firms spent a lot of energy in communicating the value of their services.

6.3 Communicating service value

Communication may significantly affect the service experience because of the very nature of high-tech services. We know that a service is an experience whose performance is measured by its ability to fulfill customer needs and expectations.

But the value of a service is very much driven by its perception from the customers, because perceived performances are much more important than actual performance.

As a matter of fact, the perception of the service can be chiefly modified by communication before, during, and after the performance. Consequently, the communication strategy is important not only to create service value, but to make it clear and to emphasize its uniqueness to the customers.

6.3.1 Communicating before the service performance

Before customers have experienced a service, the communication may modify their expectations about the service and may influence their selection of vendors as well as their decision making process.

As for any product and service, the first goal of the communication strategy is to raise the level of awareness of potential customers. A company has to make sure that the markets will know about its services and its offering. This is clearly the problem some high-tech services vendors have when they belong to a multibusiness firm like IBM Global Services or Hewlett-Packard. This was also the case for Andersen Consulting whose image for a long time was overshadowed by its mother company, Arthur Andersen, the famous accounting and finance firm, which even had its own service offering competing with Andersen Consulting. One vice president of Andersen Consulting said: "We have very aggressive communications, based on mass media in order to build our brand and to differentiate it from our direct competitors and from Arthur Andersen."

Most specifically, because high-tech services are often perceived as hazardous due to their innovative content, at that stage the communication will try to reassure potential customers and to present the vendor as being the safest and more credible. This is the reason why high-tech services vendors advertise their existing references to demonstrate their know-how with tangible users. They try to create a successful service image through proof and testimonials to convince hesitant customers by showing what they can do.

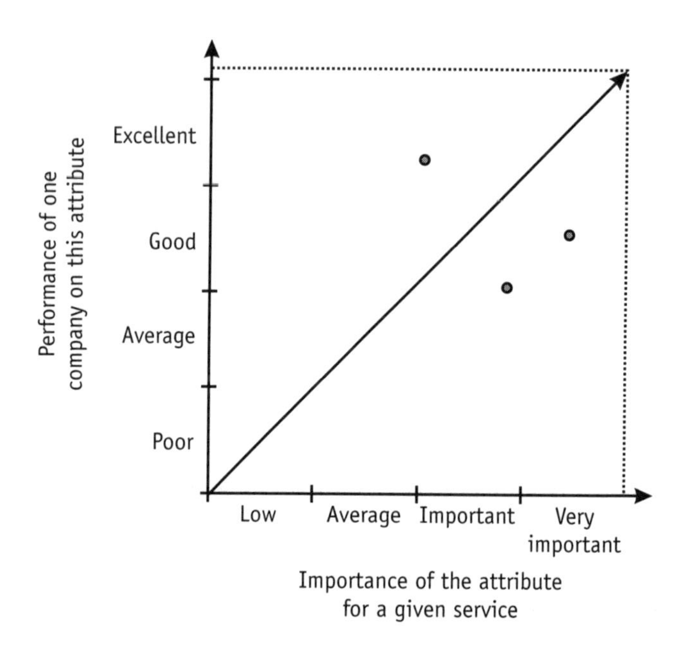

Figure 6.6 The importance/performance matrix.

The communication strategy may also influence consumers' perceptions by identifying the key attributes of a service, which really matters for customers in a given market segment. The next step is to compare the weighting of each attribute and the customers' scoring of the perceived performance of the firm (see Figure 6.6).

Then the company may intend to modify its score (or one of its competitors) on a given attribute with a specific communication campaign designed to change the customer's perception. The more important the gap between the actual performance and the customer's perception, the more effective the communication campaign.

6.3.2 Communicating during the service performance

Communication may also play a role during the service performance when the customer goes through the service experience, sometimes called the "moment of truth." This is principally the case for Web-based services because the consumer is more or less involved in the service process. Actually, to get into the service and to obtain the output sought, the consumer must perform various operations on different screens by clicking icons or entering information requested by the Web site.

In such a process the consumer is acting as a part-time employee according to a script designed by the vendor [10].

The consumers' satisfaction will be greatly dependent on their ability to successfully go through the various stages of the script. Too many screens, too long of an answer time, too much information requested, too many lengthy, boring, or complex operations (such as filling in two screen pages with personal information) will have a negative impact on the customer perception of the service. AOL won a significant number of customers by making its screens as simple and fun as possible.

The script, the *modus operandi* of the server on the Web, is a key element of the service, whatever its actual performance. Today, some netchandizing companies like Businesslab, CKS, PoppeTyson, and many others, are helping Web-based service providers to better communicate with their users through effective and attractive programs.

Regarding the service offered to the business users, most of the communication is managed directly by the account managers and the contact employees. We have seen in Chapter 5 that the behavior of these people was an essential element for convincing and reassuring the customer at various stages of a service project.

6.3.3 Communicating after the service performance

Communication may also play a role by modifying the perception of the user after the use of the service. Such is the case when the service is underperforming compared to the customer's expectation, leading to customer dissatisfaction.

Indeed, effective communication is often the key to retaining or regaining an unhappy customer. It starts with a formal process to identify unsatisfied customers, mostly by giving them the opportunities to expose their dislike through mail, phone, e-mail, or face-to-face to an explicit address and contact name.

The next step is to listen carefully to customers' complaints, an essential element of communication, before answering the problem. The problems can be technical, in which case the best solution is to explain to the customers the reasons for the problems, to apologize, and to offer free compensatory use or the monetary equivalent.

The problems may also be relational and may come from too high a level of expectation from users compared to the actual performance; in this case, a specific explanation must be provided to the dissatisfied customer, which will downplay the over-exaggerated expectations of some service attributes (for instance, the speed of displaying screens or data) while emphasizing attributes that may be under-evaluated by the client (for instance, security, which requires the use of password, ID, or encryption techniques that may slow down the availability of the service for the user).

There is no effective service recovery program without efficient communication. However, if too many clients have the same problem, it means that the solution needs to be found up-front in an effective communication campaign before and during the consumption of the service.

6.4 The communication mix for high-tech services

The communication mix includes both the communications tools available for the promotion of the high-tech services, and the way those services can be used as communication tools themselves.

6.4.1 The communication tools to promote high-tech services

Figure 6.7 lists the various communications tools that can be combined by the marketers in a communication mix for high-tech services. Those tools are not very different from the ones used by high-tech product vendors. Their use depends on the communication objectives of the company and its various target audiences.

Regarding the objectives, one must note that all communication tools do not have the same purpose with regard to stimulating a purchasing decision. Some tools are appropriate for establishing awareness; others are excellent for communicating a better understanding of a service or a technology; even others strengthen the appreciation for the service or finalize the purchase.

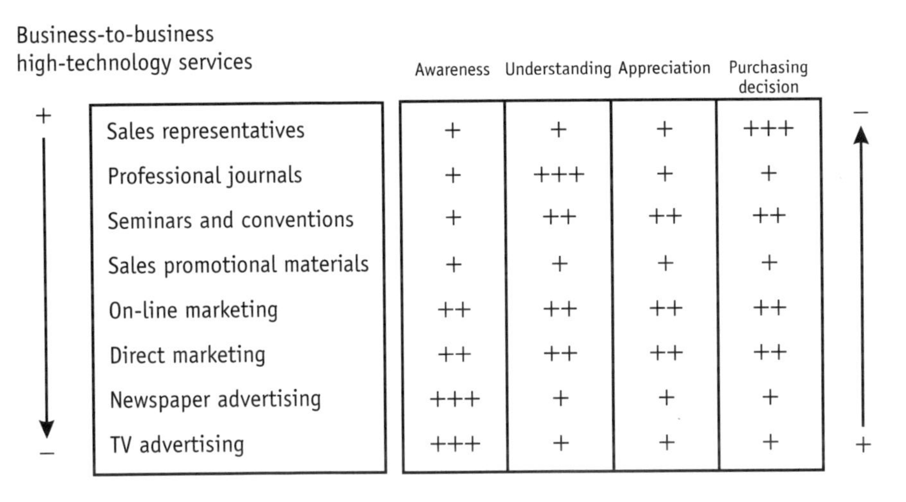

	Awareness	Understanding	Appreciation	Purchasing decision
Sales representatives	+	+	+	+++
Professional journals	+	+++	+	+
Seminars and conventions	+	++	++	++
Sales promotional materials	+	+	+	+
On-line marketing	++	++	++	++
Direct marketing	++	++	++	++
Newspaper advertising	+++	+	+	+
TV advertising	+++	+	+	+

Figure 6.7 The importance and efficiency of various communications tools for high-tech services.

The type of customer a company is targeting also drives the use of communication tools. Most specifically, the communication mix will strongly differ when talking to organizations or to consumers.

When communicating with business customers, high-tech services providers rely mostly on their account managers whose vocation is not only to sell, but also to communicate effectively with the customers. The account managers are in charge of listening to customers to identify new needs, to introduce new solutions, to answer any complaints, and to oversee various teams at work with a major customer. Accordingly, a good account manager is not only a good sales representative but also an effective communicator.

Professional magazines that specialize in on-line and electronic services are a preferred medium when presenting an innovation or developing the advantages of new services. Among the most famous are *Infoworld*, *Computerworld*, or *Service News*. These magazines feature articles written by professionals, interviews of leading specialists (including top developers from the engineering departments of high-tech services vendors), and news releases.

Advertising in trade magazines to business customers allows vendors to take advantage of the high level of credibility of these professional publications. Furthermore, communication campaigns for services can be developed using a technical angle minimizing the risk of being misunderstood, because the readers of these publications are familiar with technology.

Other communication publications are sponsored magazines, like *IBM Think,* or *EDS Government Industry Group News*, and newsletters. They are distributed to customers and interested parties at no cost. They contain useful articles, but can gain respectability even though they are clearly partisan. An executive of Syseca said: "Our communications with customers focuses on a quarterly review edited in four different languages and on specialized seminars centered on one topic for one country."

Actually, *seminars* are also educational and as communication devices, are particularly suitable for high-tech services. A new technology or a new service for a category of business can be thoroughly explained during a two or three day seminar. Customers can familiarize themselves with the technology before adopting it. A well run seminar explains what a technology or a service is about and shows that this technology or service functions well.

Sales promotional materials are mostly catalogs and service description literature that try to present the technical characteristic of each solution, emphasizing the idea that a picture can say a thousand words.

On-line marketing is heavily used by high-tech services vendors to communicate with all types of customers, consumers, or businesses. Section 6.4.2 is dedicated to this new communication medium.

Direct marketing, through mail or telephone, is not used very often to communicate with corporate customers. This is also the case with newspaper advertising and television.

These mass media are, on the contrary, the preferred tools of on-line services vendors that want to communicate with consumers. They offer a reasonable cost per thousand contacts and cover enormous target markets. They also have the ability to quickly create a high level of awareness for a brand or a service. In absolute value, mass media are expensive and usually reserved for large companies that target sizable market segments.

All in all, the communication mix of high-tech services does not differ significantly from that of more traditional services, except for one major divergence, which does not come as a surprise: on-line advertising has a much more important share because of the Internet. Therefore, presented next are some specific comments on this communication tool.

6.4.2 Publicity and advertising on the Web

Just being there on the Web is already a communication act, which can generate market awareness and positively affects the attitudes of customers for a service or a firm. According to the vice president of Syseca: "We have noticed that, in the Anglo-Saxon countries, we have been consulted on various service projects only because our Web site had been consulted through a query at the preselection phase of potential bidders." What is true for business customers is also true for consumers who are looking for new services by going through search engines like Yahoo, Excite, or AltaVista or by chatting in various newsgroups such as GeoCities.

As a rule for gaining some publicity, any new site has to be submitted to search engines, portals, and directories and needs to be registered in various mailing lists and newsgroups, as well as in the on-line yellow pages.

Nonetheless, the Internet can be used not only for publicity but also for advertising to inform and persuade customers through paid media. Not too long ago, the thought of advertising on the Internet was the farthest thing from the minds of marketing executives.

For 20 years, the old Internet held firm to its conviction that advertising and marketing held no place on the Internet—it was above all such commercial activity. But the World Wide Web and its appeal to marketing professionals have completely changed that. The Internet Advertising Bureau via its industry-leading Advertising Revenue Reporting Program, announced that $906.5 million was spent on on-line advertising in 1997. By the year 2000, Forrester Research Inc., a technology market research firm based in Cambridge, Massachusetts predicted that this figure will hit $2.2 billion.

The most effective promotional tool has been the ability of the Web to carry the messages of marketing and communication professionals to targeted groups. Here, promotional information like new product announcements, product catalogs, and training and seminar schedules, can be put on the Web. Interested customers and prospects can instantly click—get the information—and respond interactively. And if a prospect goes to the trouble of connecting to and reading a Web site, he or she is more likely to be a source of business.

But the question remains—will the right people see the information? The massive scope of the Web can leave companies lost and unnoticed—buried deep in the sea of over 100,000 sites on the Web today. The answer to the advertiser's dilemma mirrors the change that has occurred in retail marketing over the past decade as retailers have had to put their stores in malls, where millions of customers can come en masse.

A similar move is taking place on the Web today. Electronic marketplaces that focus on the interests of targeted buying groups are being developed to include companies that want to offer their services to that group. For example, in the industry sector, Online Marketplace has emerged as the largest and most-used industry-focused electronic marketplace. It currently includes over 450 of the leading manufacturers of services and products used by industry buyers and specifiers in serving their business needs.

Similarly in the consumer business, companies such as GeoCities are hosting communities of users who share similar needs, interests, and topics. The 2 million "homesteaders" at GeoCities design homepages in various "neighborhoods" and they—as well as the million people their pages attract—represent a target group of interesting users for a wide array of service providers.

What differentiates interactive media such as the Internet most from the more traditional forms of advertising (for example, television, radio, magazines, and newspapers) is that it is more buyer initiated. Consequently, on-line advertising has the potential to be significantly more effective, due to the higher interest and attentiveness levels. Furthermore, the advertiser is not limited to a page in a magazine or a 30-second time frame in which to communicate the key product benefits to a broad segment of the target market.

While banners remain the predominant advertising vehicle, a survey conducted by Coopers & Lybrand in 1998, found that sponsorship of on-line content increased notably, capturing 41% of the total revenues. Banner advertising accounted for 54% of the total.

According to some industry estimates, during 1996, some 600 differently sized advertising banners were in use on the Web. The U.S. trade bodies proposed eight standard sizes to make it easier for advertisers to create electronic ads. The proposed sizes (in pixels) are listed in Table 6.1.

The impact of banners demonstrated high levels of advertising banner awareness. The same study found a dramatic increase of awareness after a single advertising exposure, as well as a significant impact on brand perception resulting from one exposure. This study conclusively demonstrates that on-line advertising has the same communication power as traditional media.

The exposure of a single Web advertising banner enhances positive perception of advertised brands and improves the likelihood of consumer purchase. Advertising banner exposure was found to be responsible for 96% of advertising awareness, compared to click-through which contributed only 4%. This important finding indicates that banners are not just direct marketing vehicles to drive users to brand Web sites, they are powerful advertising communication vehicles as well.

The study also tested how Web users feel about on-line advertising in general. Of those surveyed, 63% strongly agreed that brands advertising

Table 6.1
Banner Type and Proposed Sizes

Banner Type	Proposed Sizes
Full banner	468 × 60
Full banner with vertical navigation bar	392 × 72
Half banner	234 × 60
Square button	125 × 125
Button #1	120 × 90
Button #2	120 × 60
Micro button	88 × 31
Vertical banner	120 × 240

on the Web are more forward thinking than other brands, and 55% of the participants were in favor of Web advertising.

There is a growing body of information about the theory and practice of advertising on the Internet. Through marketing research and theoretical cross-examination, industry experts have concluded primarily six important factors for on-line advertising (see Figure 6.8).

1. *Exposure.* The Web allows manifold exposure for messages in various sites with multiple signposts. They can be positioned adjacent to content that is relevant to the targeted audience. This is very important since messages found in context are always of more value and, therefore, are likely to be sought out and remembered.

    Service vendors or advertising agencies, such as DoubleClick, use the cookies technology to track users' browsing habits. Doing so, they get effective profiles of Web users and deliver well targeted messages to very segmented audiences.

2. *Frequency.* On the Web, message frequency is a function of user control. To make effective on-line advertising, service vendors

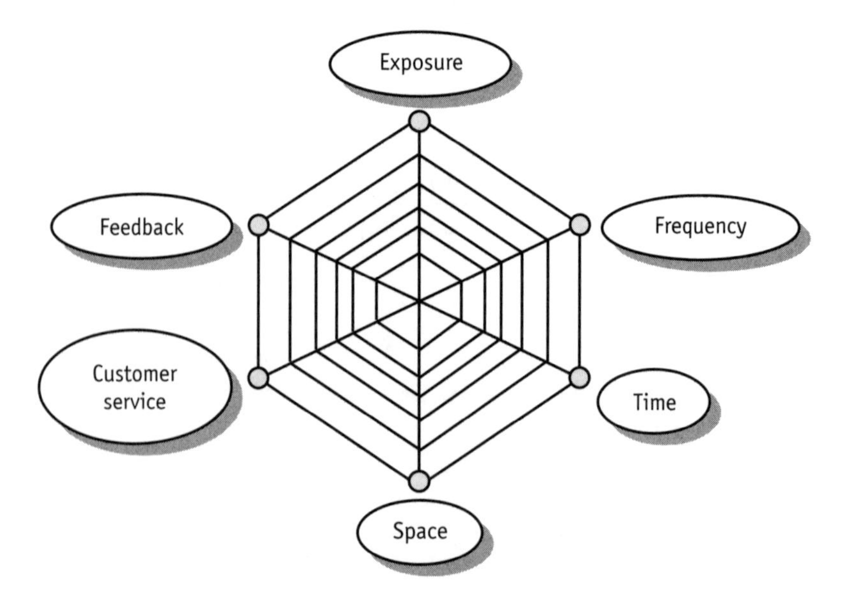

Figure 6.8 The six key success factors for advertising on the Web.

must get the user to come back by making their messages appealing, attractive, and useful.

Even without click-throughs, banner work delivers a positive effect on brand perception. The average efficiency of Web advertising is higher than television and close to that of print. All ads contribute to awareness, but the results vary significantly according to the creative execution. A good banner ad works much more effectively than a bad one. Every advertising impression is important, whether it results in a click-through or not.

3. *Time.* Time is a key dimension to be mastered in order to advertise usefully on the Web. Advertisers must be very sensitive to delivery time and how fast information is retrieved. They have to understand that a second's delay may prompt a consumer to think about going elsewhere, or worse, to think about the cost of being on line.

   One word of advice is always to remember that graphics take longer to download than text. They have to be used sparingly, where they enhance a story—not where they simply look good.

4. *Space.* Space is virtually unlimited on line. This allows service vendors to deliver raw volumes of information, a significant advantage compared to traditional communication media. Nevertheless, this information must be easily available and made accessible in layers. It should be easy for the consumer to find what he or she is looking for and skip the rest.

5. *Customer Service.* The infospace can be used not only to sell, but also to provide post-sales support because on-line users love e-mail and use it frequently to express their opinions. Given the opportunity through an e-mail address, users will tell what they think about a service and what they want. They'll also expect a response from the vendor. Provided so, Web advertising is a superb opportunity to foster relationships and to build lifetime customers.

6. *Feedback and Research.* Because the on-line user expects to be involved in a way that other consumers do not, on-line space may also be applied to collect user information and to measure

feedback for usage patterns. Advertisers have the chance to solicit feedback via surveys or electronic "suggestion boxes" to constantly refine the content and organization of their electronic ad. Such a process is better and faster than the usual post-test used to evaluate the efficiency of an advertising campaign in traditional mass media.

Besides those six elements, there is another issue which is key for the diffusion of advertising on the Internet, the matching of the lowest common denominator of a Web browser, as far as the versions and the plug-ins are concerned. Too often, users cannot access or read information (or advertising messages) on a Web site because they do not have the requested version on their computer.

6.5 The pricing policy for high-tech services

The pricing of high-tech services does not differ dramatically from more traditional services. However, there are some interesting differences giving new opportunities to marketers.

6.5.1 Pricing options

As for any product or service, the range of pricing options for high-tech services is limited between a ceiling, set by the market segment that is being addressed, and a bottom price established by the cost structure of the service. Beyond the ceiling, no customer will use the service, which means a zero market share.

Below the bottom price, the service vendor is losing money, which means a negative return. Actually, a service vendor can build a high volume infrastructure and price below cost to encourage usage; but such a move can go on for a limited amount of time because of the financial loss associated (furthermore, in many European countries, pricing below cost is forbidden by law).

Between these two limits, the offering of competitors may impact the price. Accordingly, these three price pointers translate into different pricing techniques (listed in Figure 6.9).

Price pointers	Pricing techniques
Market segment acceptance	Comparison with substitute services Pricing to value
Competitor's pricing strategy	Market price Bidding price
Service cost structure	Break-even point Cost + profit margin

Figure 6.9 The different pricing techniques.

The simplest method to make a price is to add a profit margin percentage to a total cost. In that matter, high-tech services offer a greater amount of flexibility to marketers for two reasons.

First, their cost structure is not very dependent on capacity and time, contrary to more traditional services. Indeed, airlines or restaurants have to set up prices to adapt to the demand of their existing capacity at a given time, should it be an airplane or a restaurant's dining room. Typically, pricing (often called "yield management") is set up to stimulate demand during periods of low utilization of capacity and to level off consumption peaks by moving customers to less crowded periods.

Such is not the case with high-tech services, most notably on-line services, which rely on an extensive communication infrastructure and technology that is not easily saturated (although, recently AOL was shut down for 6 hours because of traffic overload). Web sites run 24 hours a day on the Internet and customers do not need to move in a physical space to use the service, they just have to click and access. However, one must note that people-intensive services such as consulting, help desk, or software support can get more easily saturated because of the lack of people.

Secondly, the nature of the cost structure is very different from traditional services. Because digital assets are not depleted in their consumption, they can be provided endlessly almost at no cost to feed an infinite number of transactions. In other words, the variable costs associated with high-tech services is almost nil, meaning that the additional users come with a big profit, once the fixed costs (mostly computers and communications) have been amortized. High-tech services vendors can even offer a service for free in order to increase its acceptance by customers.

The economies of scale for standard goods have their equivalent in economies of knowledge for high-tech services, except that a very limited infrastructure can accommodate a very big number of users giving the opportunity to small firms to compete on the same footing with large firms. Actually the Web's power as a sales vehicle has proven to be inversely proportional to the size of the seller because the Web's worldwide reach can instantly transform a small outfit—such as Amazon.com or Auto-by-Tel—into a global distributor. By contrast, large corporations that already have their distribution networks in place—such as Marriott, Barnes & Noble, Fidelity, or Nissan—find the Web to be only a niche channel.

Obviously, economies of knowledge work better with technology intensive services, such as network services or on-line services than with people intensive services, like consulting or system integration.

Independence from capacity and time associated with a limited ratio of variable costs make it easier to break even. They also make for low prices because demand is very elastic on the Web, that is, sensitive to price variation, for there are a lot of alternatives available to the customers.

This is less the case in business-to-business marketing where demand is less flexible; some services have few substitutes because of the absence of competitors, and buyers are less sensitive to price than to additional performance.

Thus the successful high-tech services companies are pricing their service to value, whenever they are in a position to offer a unique service with no equivalent on the market. This pricing technique is based on a simple assumption that when a service answers customer needs, those customers know the value of the service by taking into account its different benefits. Consequently, they know the price that they are ready to pay for it. One executive of IBM Global Services pointed out:

> Our pricing policy is based on both the measurement of our services' added value, and our willingness to build an open and long-term relationship with our customers. In our business, there are big teams that are going to work together for a long time. If the relationship does not start well, the service process is going to fail. It is better to expose its costs and profitability constraints to the customer and to discuss them frankly.

Ultimately, many services are offered in a competitive market.

6.5.2 Revenue models

The revenue models differ strongly between services offered directly to businesses and services offered on the Internet to all types of customers.

Most of the high-tech services delivered to corporate customers are projects negotiated at a fixed price with a revision clause contract. These projects last from 1 week to more than 3 years. They are phased in as "deliverables," accepted by the customers, and "milestones," paid for by the customer, which generate the revenues for the service vendor. Each phase can be the opportunity for discussion and renegotiation.

Accordingly, a good level of reciprocal trust is a must to ensure that the whole project will go smoothly. The conditions for payment may vary because it is often linked to the value creation of the service vendor; in an extreme case, Andersen Consulting services fees are directly indexed on the number of Smart cars produced by its customer, Mercedes. But basically, the whole flow of revenues is clearly identified up front in the service contract and is more or less predictable.

Such is not the case with on-line services where customers are often unpredictable, both in terms of number and in terms of service consumption because of their intense volatility. Because of the nature of the Web and the heavy competition, users can switch easily from one site to another thanks to graphical browser interfaces.

Different revenue models have emerged for selling services in the Web marketplace:

- Charging fees for on-line transactions or links;
- Charging fees for the content accessible on a Web site;
- Selling advertising space.

These three models can be combined to maximize revenues.

6.5.2.1 Charging services

One emerging revenue model involves charging customers for transactions or any type of service, such as searching databases or providing space and additional services to clients on a Web site.

One of the most successful service providers on the Web is Industry.Net, an on-line marketplace for manufacturing and industry. Industry.Net, which started as a dial-up service, went on the Web last October, offering businesses a place to shop for manufactured products and goods. It charges manufacturers and suppliers $3,000 to $8,000 a year to maintain an "electronic storefront" on its site. In this way, Industry.Net has generated about $20 million this year in Web business alone. It expects to bill another $8 million by year's end.

In the emerging Internet search business, InfoSeek Corp. in Santa Clara, California, makes money by charging users subscription and transaction fees for its database search service—and by selling space to advertisers. Since launching in February, InfoSeek has generated about $14.5 million in advertising and subscription revenues, though the company declined to break out the figures by revenue stream.

6.5.2.2 Charging content

Another way of making money on the Web is to charge users for access to content, most commonly with subscription fees. While a few pricing models are beginning to emerge, there are no clear success strategies yet.

The biggest obstacle in selling subscriptions over the Internet is that consumers are used to getting cyberspace information for free, thanks to the dozens of free news and information Web sites that are available.

For business-to-business services, this sort of model is viable because businesses are used to paying for specialized information that will give them competitive advantages. For example, Quote.Com Inc. provides stock quotes, company profiles, annual reports, news, and other business information. Quote.Com charges $10 for its basic service, $34 a month for its premium service, and it does not accept advertising. It has generated $650,000 since launching in July 1994, and it is profitable, said company president Chris Cooper.

Most content providers charging for subscriptions also count on advertising revenues from their sites. In fact, many agree that advertising really will drive the business, but charging subscription fees will not generate significant revenue on the Web. While companies will spend $2.2 billion on Web advertising by the year 2000, Web subscriptions that year will come to only $156 million.

6.5.2.3 Selling advertising space

Some companies looking to make money from the Internet are turning to a centuries-old business model: advertising. In 1998, companies spent roughly $1 billion to advertise on the Web and that amount should double in the next two years. Sites that include HotWired and America Online Global Network Navigator charge advertisers as much as $15,000 per month to post information about their companies, products, and services. To stimulate the audience, companies like GeoCities reward Web page owners who get many hits.

No standard rate structure has emerged for the Internet (see Table 6.2). HotWired charges $15,000 a month for an ad, with discounts

Table 6.2

Some Popular Advertising Sites on the Web

Site	Weekly Visitors (source)	Monthly Rate (CPM)	Notes
Excite	1,000,000 (Proprietary)	$11,000 ($24 cpm)	Search engine
HotWired	40,000 (Proprietary)	$15,000 ($150 cpm)	On-line Publication
Pathfinder	892,000 (Nielsen)	$12,600 ($29)	On-line Publication
Yahoo	2,053,571 (Proprietary)	$20,000 ($20 cpm)	Search engine

averaging 15% for advertisers who also buy into Wired's print edition. Netscape, which controls 70% of the Web browser market and whose browser defaults to its site whenever a user launches onto the Web, charges companies $15,000 to $30,000 per month to place ads on its highly trafficked site.

One sticking point for Web ads is that there is no way to get a decent measurement of the advertisements' effectiveness. The most popular method for measuring Web usage is one that even Web experts concede is useless—"hits," or the number of files requested on a site. The problem is that hits have little or no correlation with the actual number of visitors to the site. For instance, one user accessing one Web page will score five hits (not one) if the page includes five graphics icons stored as separate graphic files.

Market researchers, such as Intersé and NetCount, have developed technologies that can differentiate hits from real visits. They provide more realistic measurements of how much interest an ad on the Internet generates and could one day become the Nielsens or the Burkes of the Web.

6.6 Chapter summary

The "marketing-mix" for high-tech services is slightly different than for product or traditional services. First, the distribution strategy is very limited.

Regarding the conception of new services, it is interesting to notice that high-tech services allow the development of solutions in line with the principles of relationship marketing. The ultimate goal of relationship marketing is the one-to-one marketing where each segment of the market is only one customer, meaning that the firm is able to deliver a specific solution to each of its customers. This is clearly the case for professional high-tech services because of the limited number of clients. But it also works for on-line services to consumers since it is easy to interact directly with users on the Web and to tailor the offer to the profile of each customer.

High-tech services offer flexibility, which makes them more easily customizable than traditional services. They fit very well with the model of mass-customization, which aims at satisfying each customer while enjoying the economies of scale of mass production. Like high-tech products, they also include a significant amount of technology, which means that whether some new services are clearly market driven, others are pushed by the technology.

Whatever the origins of a new service—in house or from customers—all leading high-tech services firms have roughly the same development process with a concept test, followed by a feasibility study, then the design phase, succeeded by the development and testing before the final release.

The communication strategy is very important not only because what is sold is intangible but also because the service can have a significant impact on the activity of a customer. The communication strategy includes brand image management, effective communication of the service value to customers, and the precise definition of the target audience.

In the high-tech services business, branding is an absolute necessity for many reasons:

- It changes the essence of the service and adds a psychological value.
- It helps customers to make decisions.
- It facilitates identification.
- It reassures customers.
- It contributes to differentiation.
- It justifies a price premium.
- It boosts staff morale.
- It eases recruitment.

A strong branding strategy is based on three key principles—dominance, exclusivity, and singularity—and requires a significant amount of money to promote the brand up front.

Most of the successful high-tech services companies have made the decision to promote their corporate brand name, using it as an "umbrella" for the various categories of services they are offering. They start with the firm identity defining the brand identity before designing the instruments to show this identity. Ultimately, a strong brand name comes from matching the image set up by the firm with customer experience.

But the value of a service is very much driven by its perception from the customers, because perceived performance is much more important than actual performance. As a matter of fact, the perception of the service can be chiefly modified by communication before, during, and after the performance.

Regarding the communication mix, this chapter details both the communication tools available for the promotion of the high-tech services and the way those services can be used as communication tools themselves, most specifically on the Web.

The pricing of high-tech services does not differ dramatically from more traditional services. However, there are some interesting differences, which give new opportunities to marketers. First, their cost structure is not very dependent on capacity and time, contrary to more traditional services.

Secondly, the nature of the cost structure is very different from traditional services: since digital assets are not depleted in their consumption, they can be provided endlessly almost at no cost and can feed an infinite number of transactions. This economy of knowledge gives small firms the opportunity to compete on the same footing with large firms.

Independence from capacity and time associated with a limited ratio of variable costs makes it easier to break even and also makes for low prices on the Web where demand is very elastic. This is less the case in business-to-business marketing where demand is less sensitive to price variation. Furthermore, successful high-tech services companies are pricing their service to value, whenever they are in a position to offer a unique service with no equivalent on the market.

Regarding the revenue models, they differ strongly between services offered directly to businesses, and services offered on the Internet to any type of customer. Most of the high-tech services delivered to corporate customers are projects negotiated at a fixed price with a revision clause contract.

Such is not the case with on-line services where three different revenue models have emerged, namely, charging fees for on-line transactions, for the content accessible on a Web site, or for selling advertising space. These three models can be combined to maximize revenues.

References

[1] Dibb S., et al., *Marketing, Concepts, and Strategies*, Boston: Houghton Mifflin Company, 1994.

[2] Bennet, P. D., ed. *The Dictionary of Marketing Terms,* Chicago: American Marketing Association, 1988.

[3] Christopher, M., A. Payne, and D. Ballantine, *Relationship Marketing*, Oxford: Butterworth-Heinemann, 1991.

[4] Vavra, V. G., *Aftermarketing: How to Keep Customers for Life Through Relationship Marketing*, Homewood, IL: Business One, Irwin, 1992, pp. 2–6.

[5] Ryans and Wittnik, 1997, p. 314.

[6] Woods, T., and Remondi J., "Relationship vitals for high tech marketers," *Marketing News,* Vol. 30, May 20, 1996, pp. 8–9.

[7] Von Hippel, E., *The Sources of Innovation*, New York: Oxford University Press, 1988.

[8] Knowles, A., "Find out what your customers really want," *Datamation,* Vol. 43, No. 2, February 1997, pp. 68–72.

[9] Tocquer, G., and C. Cudennec, "How the Tigers Got Their Stripes," Hong Kong: Trade Media Holdings Limited, 1997, p. 126.

[10] Soloman, M. R., et al. "A Role Theory Perspective on Dyadic Interactions: The Service Encounter," *Journal Of Marketing,* Vol. 1, No. 49, 1985, pp. 99–111.

7

The Challenges Ahead For High-Tech Services

THE IRRESISTIBLE GROWTH of high-tech services both for consumers and business customers seems to experience no barriers for the time being. We have seen that successful firms are the ones that balance strong customer focus with mastering leading edge technology, as well as sound management of their people. Is that enough to sustain a continuing expansion? No, whatever market they serve—businesses or consumers—high-tech services firms must keep pace with the customers' constantly changing environments in order to be successful.

An executive of Andersen Consulting said, "Our business changes permanently because of the evolution of our customers' needs. Five years ago, we were doing a lot of consulting; today we are working on large projects including outsourcing services which are increasing." This can be explained by a major turn around in the attitude of corporate customers. In the 1980s, they wanted to manage the risk of integrating various technology themselves. But since the beginning of the 1990s, they have moved to outsourcing the integration of information-based technology and other services for various reasons:

- Their own customers are changing so fast that in order to adapt they must focus their resources on their core competencies and outsource all the less important components of their value chain.
- Technology is moving so briskly that no one—even large companies—can or wants to keep up with the pace, except for high-tech services firms, which have both the competence, the knowledge, and the volume of activity to hedge the risks.
- The costs of developing an in-house solution—which may fail, be quickly outdated, or too costly to maintain—are too high compared to the use of an external service offering the same level of performance.

Interestingly enough, consumers are turning to high-tech services for similar reasons. They are looking for services that help them keep up with the changes as shown for instance, by the success of e-mail or chat services. At the same time, they do not want to be bothered by the hassles of technology (as AOL figured out first and has since become the indisputable leader of on-line services). Ultimately, consumers look at reasonable costs such as the ones offered on the Web. Consequently, high-tech services vendors must constantly redefine themselves. One Atos executive speaks for many colleagues and competitors when remarking: "The future of companies like ours lies in the anticipation of customers' needs, the ability to answer them with complete solutions, and the capacity not to be only a supplier, but to provide them with a vision." Indeed, high-tech services must constantly bring a real value addition to the customers, consumers, or organizations. They must offer them "future sourcing."

About the Author

Eric Viardot has a Ph.D. in management. He is a graduate of the HEC Business School, Paris, and the Institute of Political Sciences, Paris. After working for Digital Equipment, Dr. Viardot was a strategic consultant at Bain & Company. He is now a professor of marketing and strategy at Ceram Graduate Management Business School in Sophia Antipolis, France. He frequently advises general management in strategic and marketing decisions.

Index

Recent Titles in the Artech House Technology Management and Professional Development Library

Bruce Elbert, Series Editor

Designing the Networked Enterprise, Igor Hawryszkiewycz

Evaluation of R&D Processes: Effectiveness Through Measurements, Lynn W. Ellis

Decision Making for Technology Excutives: Using Multiple Perspectives to Improve Performance, Harold A. Linstone

Introduction to Information-Based High-Tech Services, Eric Viardot

Introduction to Innovation and Technology Transfer, Ian Cooke, Paul Mayes

Managing Engineers and Technical Employees: How to Attract, Motivate, and Retain Excellent People, Douglas M. Soat

Managing Virtual Teams: Practical Techniques for High-Technology Project Managers, Martha Haywood

Successful Marketing Strategy for High-Tech Firms, Second Edition, Eric Viardot

Successful Proposal Strategies for Small Businesses: Winning Government, Private Sector, and International Contracts, Robert S. Frey

The New High-Tech Manager: Six Rules for Success in Changing Times, Kenneth Durham and Bruce Kennedy

For further information on these and other Artech House titles, including previously considered out-of-print books now available through our In-Print-Forever® (IPF®) program, contact:

Artech House
685 Canton Street
Norwood, MA 02062
Phone: 781-769-9750
Fax: 781-769-6334
e-mail: artech@artechhouse.com

Artech House
46 Gillingham Street
London SW1V 1AH UK
Phone: +44 (0)171-973-8077
Fax: +44 (0)171-630-0166
e-mail: artech-uk@artechhouse.com

Find us on the World Wide Web at:
www.artechhouse.com